Private Equity

Value Creation Analysis

Private Equity

Value Creation Analysis

A Technical Handbook for the Analysis of Private Companies and Portfolios

Volume I: Theory

Michael David Reinard

First edition. Online edition first published November 2020 on the author's private equity mathematics website (www.auxiliamath.com). Hardcover edition first published January 2021. ebook edition first published February 2021. Paperback edition first published March 2021.

ISBN: 978-1-7360778-0-1 (Hardcover)
ISBN: 978-1-7360778-2-5 (Paperback)
ISBN: 978-1-7360778-1-8 (ebook)

AUXILIA Press
P.O. Box 483
Reading, PA 19607

www.auxiliapress.com

To my dear wife Oksana, thank you for your love, support, and patience,

and to our children, Juliana, Kyrill, and Maria.

Regarding the paperback edition:

The author believes that the ideal format for *Private Equity Value Creation Analysis* is hardcover. It is designed to function as a technical handbook, and its tables of equations and data should serve as convenient references to analysts seeking to build and validate their own value creation models. Therefore, a format that is at least as sturdy as a university textbook would be preferred. The circumstances of the year 2020, however, have led to significant delays in hardcover printings. Consequently, this paperback edition is offered as a compromise for those occasions when printers cannot satisfy hardcover demand.

Contents

Preface

This is the book that I wished was on my shelf ten years ago. I was working with a private equity fund manager at the time, and one of their prospective investors, a Fund of Funds, asked us to explain what drove the firm's historical investment returns. This is usually a favorite topic for fund managers who like to talk up their companies and spin yarns about how they source deals, improve businesses, and cleverly position them for sale to strategic acquirors. In this case, however, the Fund of Funds wanted those yarns to be precisely quantified, in the form of a "value creation analysis".

We had a basic idea of how the analysis should work. Investment returns would be expressed as functions of company P&L and balance sheet changes. However, analysts have flexibility in their choices of equations and variables, and we found results to be highly sensitive to those choices. Outcomes were volatile and sometimes made little sense. Nothing in the finance literature or on the web explained the "right way" to do a value creation analysis, and the Fund of Funds was either unwilling or unable to explain how they wanted it performed. In the end, we settled on something that produced sensible results for the five or six most important investments, which satisfied the prospective investor.

The following years provided more occasions to tinker with value creation models as I worked with other fund managers and investors. Applying the formulas across different portfolios revealed which approaches were least sensitive to input variations. Systemic biases were identified and corrected by changing the way certain input variables were formatted. When I occasionally gained insight into how sophisticated global asset managers performed their analyses, I learned that not only did they all do it differently, but they generally struggled with the same issues that I was trying to solve.

For the last 15 years, I have worked with private equity and venture capital fund managers at the intersection of what institutional investors call manager selection and due diligence, and what fund managers call marketing, fundraising, and investor relations. I could see that the institutional investors appreciated value creation analysis, but since there was no "standard", they would be flexible about how analyses were performed. There seemed to be an opportunity to master the technique and give my fund manager clients an edge in their marketing efforts.

However, the mathematics needed to be rock solid. The methodologies would be scrutinized by institutions with tens or hundreds of billions under management. Thus, a proof was derived for every formula and each was tested across various capital structure and return scenarios. These efforts revealed new relationships and angles of investigation. Over time, my energy shifted from defending a specific analysis approach to figuring out how different formulas were interconnected and finding innovative ways to evaluate private companies and portfolios.

In 2019, I built a private equity math website (www.auxiliamath.com) to experiment with value creation models and solicit feedback from fund managers and institutional investors. I was encouraged to focus on two areas where there seemed to be disagreement about the "correct" analytical framework. The first was measuring the impact of leverage on private company equity returns, and the second was adjusting company P&Ls and valuations for the impact of broader market or industry trends. By the time these issues were resolved, there were over 100 different formulas. The information was too complex and interconnected to describe in blog posts, so I decided to write *Private Equity Value Creation Analysis*.

This book is exceptionally narrow in its scope. It explains changes in private company equity values, and it is written for professionals who use that information to either evaluate or market track records of private company investments. Readers are presumed to be practitioners who are associated with:

- Public or private pension funds, banks, insurance companies, foundations, endowments, government agencies, asset managers, or consultants that seek to evaluate the performance of investment managers below.
- Private equity fund managers, fundless sponsors, family offices, or institutional investors that seek to explain their performance to asset managers above.
- Academics or regulators that seek to investigate the flow of capital between these two groups, which is on the order of hundreds of billions every year.

Private Equity Value Creation Analysis describes two dozen ways to interpret and discuss private company and portfolio returns. They include: P&L-driven changes related to EBITDA, revenue, and margins; capital structure-driven changes related to valuation, leverage, and ownership percentages; and market-driven changes that can influence various items on company P&Ls and balance sheets. Each chapter highlights two or more value creation components and includes thorough derivations, examples, and interpretations of results.

This volume describes four ways of measuring value creation and three models for deriving each value creation component. Thus, the 24 different components, measured four different ways, and derived with three different models, provide 288 distinct component-measure-model combinations. Their derivations make up much of this book. In practice, only five or ten formulas are needed after settling on a value creation framework, but each alternative is thoroughly described to prove that the formulas are mathematically rigorous, and the results can be converted back and forth between different models and measures with no loss of data integrity. Those not interested in proofs may skip to the value creation "cheat sheets" in Chapter 2, which organize these formulas by model and measure.

Each chapter includes at least two step-by-step value creation examples using input variables for a hypothetical company, and results are provided for every component-measure-model combination. These are intended to help analysts build and error-check their own value creation models. More exhaustive demonstrations are provided for some of the more complicated or interesting topics, including:

- Measuring the value creation impact of leverage (Chapter 5).
- Building market proxy data sets to measure market-driven return components from proprietary and publicly available data (Chapter 12).
- Addressing holding period events like follow-on equity and debt investments, add-on acquisitions, equity distributions, and dividend recapitalizations (Chapter 13).
- Using the value creation framework to calculate Net Returns for private company investments and portfolios, including Net IRRs, without time series data or complicated cash flow models (Chapter 14).

Several chapters also include an "optional" section that is nonessential to understanding the overall analytical framework. These sometimes provide deeper explanations of underlying mathematics or suggest other angles of investigation that are more likely to prove useful in specialized circumstances, like a case study, rather than a systematic framework for evaluating many companies.

As suggested by the subtitle, *Volume I: Theory* is principally dedicated to deriving and describing every value creation formula and making the case that the analytical framework described by *Private Equity Value Creation Analysis* is more robust, more reliable, and easier to use than many alternative techniques. Its companion, *Volume II: Applications*, will provide additional examples of analyses across various transaction types and industries. It will also help analysts gather and correctly format input variables for numerous scenarios that are likely to be encountered in the real world. See Chapter 16 for a preview of *Volume II*.

Many of the topics covered in *Volume I* and *Volume II* are also available on my private equity mathematics website, AUXILIA Mathematica. The site includes a private forum and online calculators that perform many of the calculations described in the following chapters. Furthermore, anyone that purchases this book can download supporting spreadsheet files that include value creation input data, formulas, and graphs, which could be used as templates for their own analyses. Registration is free and allows users to comment and communicate with each other anonymously, while keeping spammers, scammers, and other internet ne'er-do-wells off the site. Sign up below:

https://auxiliamath.com/register

It may be helpful to explain what this book does not provide:

- It is not an introduction to private equity. Readers should be familiar with the asset class and its jargon. Otherwise, they may be misled by omitted qualifiers and oversimplifications that are made for the sake of brevity. For example, all private company investors are described as GPs, even though many are not structured as general partnerships.
- It is not about valuation. Look to the latest McKinsey & Co. edition to determine what a company is worth. Chapter 10 describes ways to flag companies that might be overvalued or undervalued, but value creation models generally assume that valuations already exist and are appropriate.
- It is not about predicting the future. Finance gets itself into trouble when it tries to be physics and predict things. This book borrows a few ideas and techniques from physics but limits them to explaining the past.

Private equity investing involves opacity. Performance and valuation metrics may be of dubious quality since private companies usually lack standardized reporting practices, significant finance teams, and actively traded shares. Data exists in silos, jealously guarded by the institutional investors and fund managers who regard their methods and hard-earned data as proprietary trade secrets. Institutional investors have limited access to company performance and valuation metrics, and fund managers lack visibility into how that data is used. This creates informational asymmetry. *Private Equity Value Creation Analysis* seeks to mitigate this asymmetry by describing a comprehensive value creation framework that is mathematically rigorous, relies on input variables that are familiar and accessible, and provides sensible and non-volatile results over the widest range of company, industry, and market conditions.

Every year, several hundred billion in private capital flows from institutional investors to fund managers, and then into private company investments, where it hopes to generate outsized equity returns and then make the return trip. While capital generally flows toward the best historical track records, it is also influenced by the quality of the stories that fund managers and institutional investors tell each other about why companies succeed, why others struggle, and how broader market or industry trends impact portfolios. Value creation analysis can add quantitative insights to these qualitative discussions, making fund managers better story tellers and helping institutional investors validate whether those narratives are supported by the numbers.

Michael David Reinard
September 2020

VOLUME I: THEORY

1. Private Equity and Value Creation Analysis

Private equity is a global multi-trillion-dollar asset class devoted to the buying and selling of private companies. It is an exciting area of finance that is often mythologized, and sometimes villainized, because its transactions are complex, its fees are high, and its assets are inherently "private". Private equity cash flow and return characteristics are unlike traditional asset classes and require various specialized measurement techniques. This book focuses on one of them, defined as value creation analysis, which can provide effective ways to interpret and discuss private company and portfolio returns.

1.1. General Partners, Limited Partners, and Funds

Private company portfolios are built by investment managers defined as General Partners, or "GPs". They may be private equity fund managers, fundless sponsors, family offices, or institutional investors. GPs invest in private companies with the goal of growing equity value and "exiting" by selling equity stakes to other buyers. GPs build private company portfolios with capital provided by institutional investors defined as Limited Partners, or "LPs". They may be public or private pension funds, banks, insurance companies, foundations, endowments, government agencies, asset managers, or consultants.

LPs typically commit capital to GPs in the form of blind pools termed "Funds". Funds help LPs to diversify portfolios, reduce portfolio volatility, and drive higher absolute and risk-adjusted returns. In practice, dozens of LPs will commit to each GP's Fund, which may invest in a dozen private companies. GPs build Funds, which are portfolios of private companies. LPs build portfolios of Funds, which are portfolios of portfolios of private companies.

Most Funds charge LPs two-and-twenty. GPs earn an annual "Management Fee" equal to 2% of Fund assets and "Carried Interest" equal to 20% of Fund net profits. Fund returns must justify these relatively high fees. They must also account for illiquidity. Funds have 10+ year lifespans and since LPs are usually stuck with them for the entire term, Fund investments are called "commitments".

1.2. Private Equity Fund Market

Globally, private equity GPs raise 350 to 450 new Funds each year. According to the Pitchbook Private Fund Strategies Report, private equity Funds raised about $500 billion in 2019, which was an all-time high (Pitchbook 2020). Preqin reported that 87% of LPs felt that their private equity portfolios met or exceeded their expectations in 2019, and 86% of LPs were planning to maintain or increase their private equity investing pace over the following 12 months (Preqin 2020).

The asset class has performed well. Top-quartile Funds raised in 2015 and 2016 have delivered Net Internal Rates of Return (IRRs) over 23.0% and 25.9%, respectively. The median Net IRR for all Funds raised between 2007 and 2016 is 14.9% (Preqin 2020). To post these results, however, LPs must consistently pick those Funds that ultimately perform better than 75% or 50% of other Funds raised, which is a feat of skill, or luck, described as manager selection.

1.3. Manager Selection Challenges

Private company portfolio cash flow and return characteristics are unlike traditional asset classes. LPs contribute capital to Funds irregularly, when GPs make new investments. Later, LPs receive distributions when investments are exited. Fund cash flows are, therefore, sporadic and not directly comparable to common market indices.

It may take a GP three to five years to build a Fund's private company portfolio. Each company may take three to seven years to work through the GP's attempted value creation plan and exit. This makes Fund returns flat or negative for several years before they accelerate, which is a phenomenon termed the "J-Curve". J-Curves complicate comparisons of Funds with different time horizons or investment paces.

Private company investing involves opacity. Unlike public companies, they lack standardized reporting practices, significant finance teams, and actively traded shares with transparent valuations. Values might be determined quarterly by a small group of GP investment professionals working directly with each company. Underlying P&Ls and other data may be unreliable, especially when key elements of GP value creation plans require professionalizing company finance functions and management teams. Unrealized company valuations are usually materially different from proceeds ultimately realized at exit.

Thus, irregular cash flows make it hard to measure Fund returns against traditional market benchmarks, J-Curves make it hard to compare them to other Funds, and opacity makes it hard to tell if non-exited company valuations are reasonable. Despite these challenges, LPs endeavor to select Funds that will generally perform better than 75% or 50% of other Funds raised and outpace the "market" enough to justify illiquidity and a two-and-twenty fee structure.

1.4. Vintage Comparisons

One approach is to assume that "performance is persistent" and GPs with historical outperformance will continue to outperform. LPs compare a GP's Fund returns to other Funds raised the same year, which is defined as a Fund's "vintage". Funds with identical vintages experience similar economic conditions and are closer to one another in their deployment lifecycles, so vintage comparisons help to normalize differences in market environments and J-Curves.

Vintage comparisons are necessary but insufficient, however, because Funds managed by GPs with long track records of top-quartile returns are probably oversubscribed. GPs raise subsequent Funds when their latest Fund is only 70-80% deployed, largely unrealized, and early in its J-Curve. LPs may not know if Fund III outperforms its vintage peers until the GP is raising Fund V. Vintage comparisons, therefore, have limited utility when evaluating new GPs or those with material team or strategy changes.

1.5. Public Market Equivalence

Rising tides raise all ships, and Funds must outperform these tides to compensate for higher fees and illiquidity. Annualized market index returns, which are common benchmarks for other asset classes, are not directly comparable to Fund return measurements like IRR. A clever solution is Public Market Equivalence, or "PME". First formalized by Long and Nickels in 1993 (Long and Nickels 1996), PME has become a common technique with numerous derivations (PME+, mPME, Kaplan Scholar PME, and others). They all provide some measure of the opportunity cost of Fund cash flows. PME is the hypothetical result of buying and selling market index shares every time a Fund calls capital or makes distributions. It is the S&P 500 or Russell 3000 but with irregular cash flows like a Fund, so Fund PMEs are directly comparable to Fund IRRs. Funds IRRs that outperform their PMEs beat the "market". The degree of outperformance or underperformance can be measured by the difference between the Fund and PME returns, or the "PME Spread".

Unlike vintage comparisons, PME Spreads can compare Funds with different investment horizons. Consider Fund A, investing through 2006-2009, and Fund B, investing through 2010-2013. Fund B may have better absolute returns because Fund A invests through a tougher environment. PME Spreads adjust for market conditions and show which Funds have better market-adjusted returns, potentially indicating a better investment strategy or greater investment manager execution.

Table 1.1 A comparison of PME Spreads

	Vintage	Fund IRR	PME IRR	PME Spread
Fund A	2006	12.6%	4.8%	+7.8%
Fund B	2010	17.7%	14.3%	+3.4%

Broad market indices like the S&P 500 and Russell 3000 are not the only options. Funds often invest in specific industries and geographies. Thanks to mutual fund and ETF managers, there are decades of available data for countless public equity strategies. An index can be found to evaluate most Fund strategies, whether they invest in software or Spain. Furthermore, proprietary market benchmarks can be built with market intelligence platforms like S&P CapitalIQ.

Thus, PME provides a market-adjusted Fund return measurement and helps to compare Funds with different time-horizons. It does not account for different J-Curves or provide any information about the reasonableness of a Fund's private company valuations. It works best when evaluating established GPs with largely realized Funds. Like vintage comparisons, PME has limited utility when evaluating new GPs or those with material team or strategy changes.

1.6. Fund Returns versus Portfolio Returns

Fund cash flows account for Management Fees and Fund Expenses (collectively termed "Fees & Expenses"), Carried Interest, and any GP capital invested in the Fund. They define a Fund's Net Return, which is ultimately the most important number to an LP.

However, a GP's proficiency at its "day job", which is the buying and selling of private companies, is best measured by a Gross Return that excludes the Fund-level items above. For example, consider Fund C and Fund D, each with a 20% Gross IRR. The former has a higher Net IRR because the latter charges its LPs higher fees. Gross Returns, therefore, suggest identical investment acumen. Fund C's higher Net Return is driven by its GP being less successful in the negotiation of Fund terms.

Table 1.2 A comparison of Gross and Net Returns

	Vintage	Gross IRR	Net IRR	Spread
Fund C	2010	20.0%	17.5%	+2.5%
Fund D	2010	20.0%	15.0%	+5.0%

Since most Funds charge two-and-twenty, or close to it, Net Returns for different Funds usually are comparable. The advantage of Gross Returns is that they are scalable. Each individual investment has a Gross Return. Combining all portfolio investments provides the Fund's Gross Return. Groups of investments, made within a Fund or across several Funds, have their own Gross Returns. Deconstructing Gross Returns by industry or investment professional is a key part of evaluating new GPs or those with material team or strategy changes.

Net Returns lack this flexibility because it is often impossible to precisely connect movements of LP capital in and out of Funds with Fund capital in and out of portfolios. The Fund's balance sheet, use of credit facilities, different LP terms and capital call requirements, recycling, short term portfolio investments, and other factors obscure the connections between Fund-level Net Returns and portfolio-level Gross Returns.

1.7. Value Creation Analysis

Alternatives to Fund-level analyses, like PME, seek to build from the bottom-up. They start with company P&Ls and balance sheets. How have they changed over the hold? How have market conditions changed? How would these changes impact returns? This is sometimes described as a value bridge or value gap analysis because it "bridges the gap" between invested and realized capital. It is also termed value creation attribution because it attributes return to various causes, often called value drivers or value levers. These volumes use the term "value creation analysis", with value creation components defined as either "return measurements" or "value drivers". An example is provided in Figure 1.1:

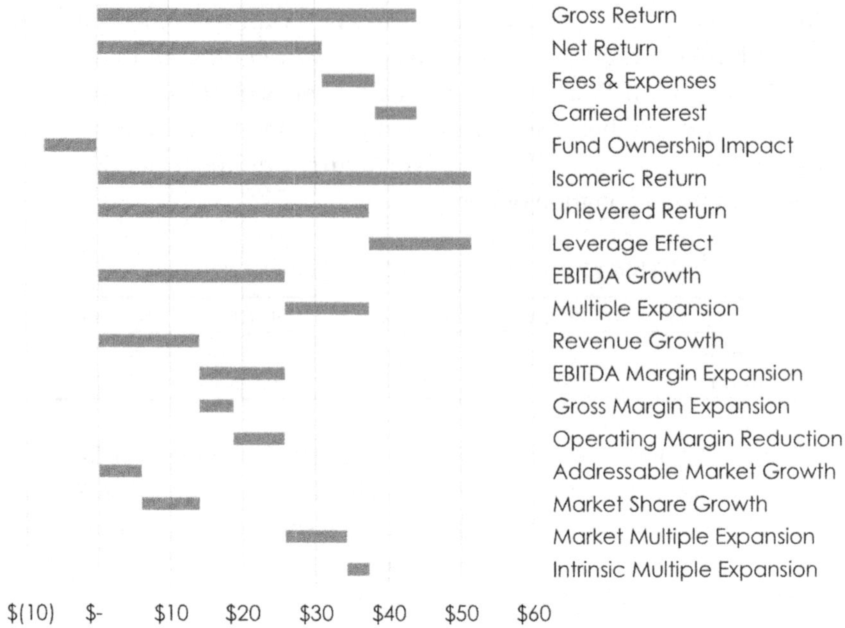

Figure 1.1 Illustrative value creation analysis

The analysis represents a Fund investment of $44 that grows to $88, resulting in Fund-level value creation of $44, a Gross Multiple of Invested Capital (MOIC) of 2.0x, and a Gross IRR of 18.9%. The figure describes 18 value creation components, including four return measurements and 14 value drivers.

- **Gross Return** is the Fund-level investment return universally found in GP reporting materials and pitchbooks. It accounts for any changes in Fund ownership that may have occurred during the hold. It excludes Fees & Expenses and Carried Interest. It also excludes the return of other shareholders, like co-investors and management, whose returns may be higher or lower than the Fund's.
- **Net Return** is an estimate of the Fund-level investment return after Fees & Expenses and Carried Interest. This measure, Fees & Expenses, and Carried Interest sum to Gross Return (Chapter 14).
- **Fees & Expenses** are an estimate of the investment's pro-rata share of Management Fees and Fund Expenses. This measure, Net Return, and Carried Interest sum to Gross Return (Chapter 14).

- **Carried Interest** is an estimate of the investment's share of Carried Interest. It is a conditional value: zero until investment value exceeds cost and Fees & Expenses; a percentage of the investment's net profits thereafter. This measure, Net Return, and Fees & Expenses sum to Gross Return (Chapter 14).
- **Fund Ownership Impact** is the Fund return driven by changes in a Fund's share of company equity. Negative numbers indicate dilution, perhaps from an expanding management option pool. Certain ownership structures, like participating preferred securities, can increase a Fund's share and generate positive values. This measure and Isomeric Return sum to Gross Return (Chapter 7).
- **Isomeric Return** is the company equity return for all shareholders scaled to the Fund's investment size. It excludes Fund Ownership Impact, which can be different for different shareholders. Isomeric is from the Greek ισο-μεϱος, meaning "equal-share". This measure and Fund Ownership Impact sum to Gross Return (Chapter 7).
- **Unlevered Return** is the hypothetical return if an investment is funded entirely with equity, or, depending on the methodology, if there are no changes in company debt levels over the hold. This measure and the Leverage Effect sum to Isomeric Return (Chapter 5).
- **Leverage Effect** is the Fund return driven by debt in company capital structures or, depending on the methodology, changes in company debt levels. It is an important measure since higher debt correlates with higher risk, which impacts LP portfolio management decisions. This measure and the Unlevered Return sum to Isomeric Return (Chapter 5).
- **EBITDA Growth** is the Fund return driven by changes in company EBITDA. GPs often get credit for EBITDA Growth, but sources could be market-driven, company-driven, or both. Changing revenue and margins both impact EBITDA Growth. This measure and Multiple Expansion sum to the Unlevered Return (Chapter 6).
- **Multiple Expansion** is the Fund return driven by changes in company valuation, which is generally a multiple of enterprise valuation divided by EBITDA. GPs often do not get credit for Multiple Expansion, but the source could be market-driven, company-driven, or both. This measure and EBITDA Growth sum to the Unlevered Return (Chapter 6).
- **Revenue Growth** is the Fund return driven by changes in company revenue. The source of growth could be market-driven, company-

driven, or both. This measure and EBITDA Margin Expansion sum to EBITDA Growth (Chapter 8).

- **EBITDA Margin Expansion** is the Fund return driven by changes in company EBITDA margin, which is EBITDA divided by revenue. GPs generally get credit for improving EBITDA margins. This measure and Revenue Growth sum to EBITDA Growth (Chapter 8).
- **Gross Margin Expansion** is the Fund return driven by changes in company gross profit margins, which is gross profit divided by revenue. GPs generally get credit for improving gross profit margins. This measure and Operating Margin Reduction sum to EBITDA Margin Expansion (Chapter 9).
- **Operating Margin Reduction** is the Fund return driven by changes in company operating expense margins, which is SG&A divided by revenue. GPs generally get credit for reducing operating expense margins. This measure and Gross Margin Expansion sum to EBITDA Margin Expansion (Chapter 9).
- **Addressable Market Growth** is the Fund return driven by changes in company total addressable market. It is a broader market trend on which GPs have little impact. This measure and Market Share Growth sum to Revenue Growth (Chapter 11).
- **Market Share Growth** is the Fund return driven by changes in company share of the addressable market. It is a measure of sales execution for which GPs generally get credit. This measure and Addressable Market Growth sum to Revenue Growth (Chapter 11).
- **Market Multiple Expansion** is the Fund return driven by changes in prevailing market valuation multiples. It is a broader market trend on which GPs have little impact. This measure and Intrinsic Multiple Expansion sum to Multiple Expansion (Chapter 10).
- **Intrinsic Multiple Expansion** is the Fund return driven by changes in company valuation discount (or premium) to the market. A positive number may indicate that a company has improved its size, growth rate, or quality of earnings since entry, especially if the company has been exited. In such cases, GPs should generally get credit for buying well or improving the business. However, a high number for a non-exited company may suggest that a business is overvalued. This measure and Market Multiple Expansion sum to Multiple Expansion (Chapter 10).

The value creation components described in the following chapters include market drivers that function like PME, leverage-related drivers that approximate financing risks, valuation-related drivers, and other components that correspond to company operational improvements, or what might be identified as a GP's "value-add". The formulas described by *Private Equity Value Creation Analysis* allow value drivers to be broken into components, rearranged, and recombined. Such flexibility can help organizations focus on their most meaningful metrics. Ultimately, combinations should simplify analyses and allow for more robust value creation comparisons across different Funds, GPs, industries, markets, investment horizons, etc.

Consider an LP whose primary concern is risk and return. They might combine Addressable Market Growth and Market Multiple Expansion into "Market Impact". It represents a return that could have been achieved with cheaper and more liquid alternatives like index funds. Leverage Effect could serve as a proxy for financing risk. EBITDA Margin Expansion, Market Share Growth, Intrinsic Multiple Expansion, and Fund Ownership Impact could be combined as "GP Execution", providing:

Table 1.3 Risk/Return value creation model

	Absolute ($) Value Creation	Relative (%) Value Creation	IRR
GP Execution	$15.3	34.8%	6.2%
Leverage Effect	$14.2	32.2%	5.7%
Market Impact	$14.5	33.0%	5.9%
Gross Total	$44.0	100.0%	18.9%
Fees & Carry	-$13.0	-29.6%	-5.0%
Net Total	$31.0	70.4%	13.0%

It makes sense to credit GPs with Fund Ownership Impact in a Risk/Return model. The dynamics of company equity waterfalls are often structured by GPs, or at least known in advance. It is part of a Fund's investment thesis and GPs should be credited (or penalized) for the result.

Alternatively, regulators or LPs with Environmental, Social, and Governance concerns may have different perspectives. Being more focused on overall company or economic health, they might choose to group Leverage Effect together with Fund Ownership Impact, Fees & Expenses, and Carried Interest, perhaps defining it as "Financing Impact". This leaves EBITDA Margin Expansion, Market Share Growth, and Intrinsic Multiple Expansion, which may be defined as "Operational Impact", providing:

Table 1.4 Stakeholder value creation model

	Absolute ($) Net Value Creation	Relative (%) Net Value Creation	Net IRR
Operational Impact	$22.8	73.6%	9.4%
Financing Impact	-$6.4	-20.5%	-2.5%
Market Impact	$14.5	46.9%	5.9%
Net Total	$31.0	100.0%	13.0%

Value creation models are robust if they allow components to be broken up, recombined, and converted back and forth between different measures with no loss of data integrity. For example, despite the different value creation component combinations in the Risk/Return and Stakeholder models, Table 1.3 and Table 1.4 provide identical Net Totals for absolute ($) value creation and value creation IRR.

1.8. Value Creation Stories

While this all seems interesting, it looks like a lot of work. An LP with a hundred Fund commitments will own stakes in a thousand private companies. This excludes the thousands of companies owned by other GPs of interest, who might manage one of the hundred prospective Funds that an LP evaluates each year. It would be impossible to perform a complex analysis on every company of interest, so where does value creation analysis fit in?

Fund-level analyses like vintage comparisons, PME, and a dozen other sophisticated techniques not mentioned in this volume are necessary, but insufficient. Otherwise, LPs would not spend hundreds of hours every year meeting and speaking with GPs. Much of this time is spent talking about team, strategy, and market, but most of these hours are spent talking about companies. At the intersection of what GPs call fundraising and LPs call due diligence, it is the stories about companies that make Fund commitments happen. Stories are where the buzzwords like "highly differentiated" and "uniquely positioned" go from platitude to possibly true. Stories bring investment approaches to life and give GPs a chance to demonstrate passion and mastery of their craft. Value creation analysis is relevant because the stories are so important. It can determine if the stories are bull***t.

A Story about Fund II

Consider a GP raising its third Fund. Fund I has strong realized returns, but Fund II has only one significant exit. That is not uncommon if Fund II is early in its J-Curve, but it is hard to gauge whether Fund I's performance might persist through Fund II and Fund III with limited data about exits. Is the lag a matter of timing or have there been fundamental changes in the GP's market or investment approach? Of course, the GP has a story about this:

- Fund II's significant exit, AlphaCo, is highly representative of our firm's approach.
- It was acquired at a reasonable valuation and levered appropriately, improved sales and margins significantly, and was repositioned to achieve a premium exit valuation.
- The same playbook was executed in several of Fund I's best investments.
- Further, Fund II's investments in BravoCo, CharlieCo, and DeltaCo, while not yet exited, have achieved significant unrealized value by accomplishing many of the same initiatives.
- Of course, there are EchoCo and FoxtrotCo, every Fund has its losers, but lessons have been learned and such mistakes should not happen again.
- The rest of Fund II's portfolio is young, with an average hold under two years, but for the most part they are on plan and at this time we see no reason why their returns would be inconsistent with our firm's historical results.

Value creation analysis is particularly good at testing the first four claims. It should confirm AlphaCo's alleged value drivers. Comparisons with Fund I companies should validate consistency of the playbook. Analyses of BravoCo, CharlieCo, and DeltaCo should show that they are valued appropriately and demonstrate real operational improvements. While it is also possible to measure the value creation components for unsuccessful deals and younger investments, these results are usually less informative. Not every company needs to be analyzed. In this case, selecting a few representative companies in Fund I and Fund II should indicate whether the GP's story hangs together.

A Story about Fund V

Measuring returns for groups of investments within a Fund, or across different Funds, is important whenever evaluating GPs with material team or strategy changes, like one with the following story:

- Mistakes were made. We had a good run with the earlier Funds, but the market changed, and, in retrospect, we were slow to adapt.
- While GPs could formerly generate respectable returns by acquiring companies with certain margins at certain valuations, markets have become more efficient, and this is no longer the case. Today's environment requires specialization, and our firm has met the challenge.
- Building on a decade of investing in Widgetry, we have enhanced our team and operational resources to become a leading investor in the space.
- Fund V will invest exclusively in Widgetry and focus on only those companies with certain financial/operational/technological advantages.
- Investments that meet these criteria make up more than half of the Fund III and Fund IV portfolios, which, excluding the off-strategy investments, would have generated strong realized top-quartile returns for both vintages.

Many GPs hit a rough spot, successfully adapt to changing conditions, and then go on to generate successful returns. There is nothing inherently wrong with this story. Evaluating such a shift can be challenging, however, because the Widgetry investments represent a "synthetic" track record without traditional Fund returns. They span multiple Funds and represent only a portion of any Fund's invested capital. Since there never was a dedicated Widgetry Fund, there is no representative track record that includes real Fees & Expenses and Carried Interest. The same would be true of a synthetic track record limited to deals led by certain professionals at a firm or the "historical" track record for a new Fund based on deals led by investment professionals at different firms.

Value creation analysis provides a mathematically rigorous and scalable framework for breaking portfolios into components, rearranging them, and recombining them into synthetic Funds. Chapter 14 describes ways to estimate synthetic Fund Net MOICs and Net IRRs, without time series data or complicated cash flow models.

Additionally, value creation analysis can assess the GP's shift to Widgetry. In the case of Fund V, it would be important to understand how much return is driven by broader market or industry forces. What if industry trends revert to the mean? How much of the industry return could have been captured by cheaper and more liquid investment alternatives? Chapter 10, Chapter 11, and Chapter 12 outline a framework for measuring the market-driven and manager-driven components of value creation.

1.9. Value Creation Model Challenges

First, the bad news. Value creation analysis requires more types of data which, from the LP point-of-view, are less familiar and less likely to be accurate. LPs are 100% certain about the inflows and outflows of their Fund commitments that drive PME. The amounts and dates of Fund contributions and distributions correspond to wire transfers from their own bank accounts.

In contrast, value creation models may require details like company ownership percentages, debt levels, valuation multiples, revenue, EBITDA, margins, prevailing market multiples, and market share. LPs are several degrees removed from these metrics. The company P&L is probably best understood by a CFO, then by the CEO, then by the GP's deal team, then by the GP's organization as a whole, including the investor relations professional that interacts with LPs, then by the LP's consultant that monitors PE investments, then by the LP's investment committee, etc.

Informational asymmetry flows in the other direction as well. When LPs regard their analytical frameworks as trade secrets, GPs are unlikely to gain visibility into how the value creation input variables are used. For example, requests for portfolio company entry multiples, with no additional context or direction, are generally satisfied by GPs sending the lowest possible values that can be rationally defended. Many deal professionals do not realize that low entry multiples tend to shift value creation away from components associated with "operational improvement" or investment manager "skill". They can be too low if the numerator (valuation) neglects transaction fees and expenses or if the denominator (EBITDA) is not really TTM or includes aspirational adjustments for things like anticipated synergies that may never be realized. It is a common mistake that GPs should avoid.

Other value creation input variables have similar issues, and they persist because GPs lack visibility into the LP methods and the LPs lacks visibility into private company valuation and performance details. *Private Equity Value Creation Analysis* endeavors to mitigate this asymmetry by thoroughly describing a comprehensive value creation framework. These volumes seek to give GPs a better sense of how the value creation formulas and input variables are used and provide LPs with a set of best practices that improve the way that they collect, analyze, and interpret private company performance and valuation data.

1.10. Value Creation Model Advantages

Now, the good news. The bottom-up value creation model provides a framework that is more flexible and offers more angles of investigation than any other technique. The 18 return measurements and value drivers demonstrated in Figure 1.1 do not exhaust the possibilities. Another ten (or so) value creation components are described in the following chapters, and the methodologies can be extended to topics not discussed, like the impact of foreign exchange rates. Since any value driver can be broken into components, rearranged, and recombined into new return measurements, the analytical possibilities are virtually limitless.

While value creation analysis requires more types of data, it generally requires less of it. Many alternatives, like PME, require time series cash flow data. Ten years of monthly cash flows for 30 investments requires at least 3,600 spreadsheet entries. Building cash flow models is not difficult, initially, but maintaining them is a battle against entropy. They grow wider with each new investment and longer with every elapsed quarter. In contrast, a basic value creation model may require only 10 or 20 input variables per company. Furthermore, with value creation models, it is much easier for companies to be added to or removed from portfolios. Without ever touching a cash flow spreadsheet, portfolios can be broken up and recombined into new synthetic Funds with portfolio-level value creation characteristics, including Net MOICs and Net IRRs.

Lastly, value creation analysis can be designed around practical input variables that are relatively easy to obtain or estimate. Procuring cash flow data generally involves special requests to GPs or deal teams. In contrast, *Private Equity Value Creation Analysis* builds formulas from the input variables that can usually be found in Fund reporting and GP marketing materials. By minimizing the need for special requests, the number of potential analysis candidates is maximized.

1.11. Summary

Most of the time that LPs and GPs spend together involves conversations about private company investments. These exchanges include stories about why companies succeed, why others struggle, and how broader market trends impact portfolios. The discussions are usually qualitative, but value creation analysis can provide quantitative measures that either confirm or contradict these narratives with numbers.

The following chapters describe about two dozen return measurements and value drivers. Formulas are designed around practical value creation input variables that are relatively easy to obtain or estimate. Derivations are mathematically rigorous, which allow values to be converted back and forth between the various return measurements. Each chapter includes at least two step-by-step value creation examples using input variables for a hypothetical company, and then compares the results of all three value creation models to assess their biases and sensitivities.

This may be more detailed than many practitioners require or desire, but the purpose of this volume is to prove that the value creation models are valid, demonstrate that the models are robust, reliable, and easier to use than many alternative techniques, and provide analysts sufficient examples to build and error-check their own models. *Private Equity Value Creation Analysis* endeavors to reduce some of the informational asymmetry in the analysis of private companies and portfolios by giving GPs a better sense of how value creation formulas are used and providing LPs with a set of best practices that improve the way that they collect, analyze, and interpret private company performance and valuation data.

Volume I: Theory is organized as follows:

- Several value creation "cheat sheets" with general notation, definitions, and formulas for the 18 components demonstrated in Figure 1.1 are provided in Chapter 2.
- Some of the principles and best practices required to build a robust and scalable value creation framework are outlined in Chapter 3. It also describes the mathematical relationships among the four ways to measure returns: absolute ($) value creation, relative (%) value creation, equity return multiples or multipliers, and value creation IRRs.
- The "big three" value drivers related to company-level EBITDA Growth, Multiple Expansion, and Leverage are tackled in Chapter 4, Chapter 5, and Chapter 6. Chapter 4 describes the most common model of value creation analysis before Chapter 5 introduces better ways to

measure the impact of leverage. Chapter 6 formalizes the three value creation models – Conventional, Derivative, and Logarithmic – and compares their results.

- Frameworks for converting company-level value creation to returns for a specific shareholder, like a Fund, are provided in Chapter 7. Fund Ownership Impact is introduced to account for value creation driven by changes in shareholder ownership percentages (e.g., dilution).
- Fund-level value creation driven by other changes in company P&Ls is examined in Chapter 8 and Chapter 9, introducing: Revenue Growth, EBITDA Margin Expansion, Gross Margin Expansion, and Operating Margin Reduction.
- A framework for measuring the market-driven and manager-driven value creation components of Fund-level returns is outlined in Chapter 10, Chapter 11, and Chapter 12. A combination of broader market trends affecting Multiple Expansion and Revenue Growth is shown to provide a reasonable alternative to PME.
- The impact of holding period inflows and outflows on Fund-level value creation is described in Chapter 13. These include follow-on equity investments, subsequent debt investments, add-on acquisitions, equity distributions, dividend recapitalizations, and divestments.
- An analytical framework for the estimation of private company and portfolio Net Returns is introduced in Chapter 14. The model provides measures of Net MOIC and Net IRR that do not require time series data or complicated cash flow models.
- Building a robust and reliable value creation framework is more than half of the battle, but results may continue to be sensitive to the way value creation input variables are formatted. Examples and case studies across various transaction types and industries will be provided in *Volume II: Applications*. A preview of *Volume II* is provided in Chapter 16.

The next two chapters provide a set of value creation "cheat sheets" with general notation, definitions, formulas, and an outline of the basic principles and best practices required to build a robust and scalable value creation framework.

2. Value Creation Cheat Sheets

This chapter describes the general notation, definitions, and most of the value creation formulas used throughout *Private Equity Value Creation Analysis*. Return measurements and value drivers are provided for the three basic value creation models – Conventional, Derivative, and Logarithmic – detailed in the following chapters. For each model, returns and value drivers are measured in the form of absolute ($) value creation, relative (%) value creation, equity return multiples or multipliers, and value creation IRRs.

2.1. Notation and Definitions

Table 2.1 General notation

X_1, X_2, or X_3	Value of X at time t_1, t_2, or t_3
$\overline{X} = (X_1 + X_2)/2$	Average value of X_1 and X_2
$\Delta X = X_2 - X_1$	Change in X between t_1 and t_2, measured as a difference
X_2/X_1	Change in X between t_1 and t_2, measured as a multiplier
$r_X = \ln(X_2/X_1) \approx \Delta X/\overline{X}$	Change in X between t_1 and t_2, measured as a continuous change
$CAGR_X = (X_2/X_1)^{1/\Delta t} - 1$	Change in X between t_1 and t_2, measured as a Compound Annual Growth Rate (CAGR)
$\Delta t = t_2 - t_1$	Holding period, measured in years
$\Delta t' = \dfrac{\ln(MOIC)}{\ln(1 + IRR)}$	Effective Hold, measured in years

Table 2.2 Fund-level return and value driver definitions

$FEqV$	Fund equity value
$\Delta FEqV = FEqV_2 - FEqV_1$	Fund return, measured as a difference
$MOIC = FEqV_2/FEqV_1$	Fund return, measured as a Multiple of Invested Capital
$IRR \approx MOIC^{(1/\Delta t)} - 1$	Fund return, measured as a CAGR to approximate an Internal Rate of Return (IRR)
$FO = FEqV/TEqV$	Fund ownership, as share of company total equity value
$\%F = F\&E/FEqV_1$	Fees & Expenses ($F\&E$), as a share of invested capital
$\%C = CI/(\Delta FEqV - F\&E)$	Carried Interest (CI), as a share of profits after $F\&E$

Table 2.3 Company-level return and value driver definitions

$TEqV$	Total equity value
TEV	Total enterprise valuation
$ND = -NC$	Net debt or net cash
$ER = TEqV/TEV$	Equity ratio
$DR = ND/TEV$	Debt ratio
E	EBITDA
$M = TEV/E$ or TEV/R	Valuation multiple
R	Revenue
$EM = E/R$	EBITDA margin
$GM = (R - COGS)/R$	Gross profit margin
$OM = SG\&A/R = GM - EM$	Operating expense margin
$\alpha_R = (R_2/R_1)/(1 + \beta_R) - 1$	Growth in market share
$\mu = M/MM$	Valuation discount (or premium)

Table 2.4 Market-level return and value driver definitions

β_R	Growth in addressable market
MM	Market multiple
$(1+\beta_R)\cdot\frac{MM_2}{MM_1}$	Market MOIC
MEM	Market EBITDA margin
MGM	Market gross profit margin
$MOM = MGM - MEM$	Market operating expense margin

2.2. Conventional Model of Value Creation

Conventional Model returns and value drivers are calculated first as absolute ($) value creation ($V$), and then converted into relative (%) value creation (υ), equity return multiples or multipliers (χ), and value creation IRRs.

Table 2.5 Conventional model absolute ($) returns and value drivers

Gross Return (V_{GROSS})	$\Delta FEqV$
Fund Ownership Impact (V_{FO})	$\Delta FO \cdot \overline{TEqV}$
Isomeric Return (V_{ISO})	$\Delta TEqV \cdot \overline{FO}$
Unlevered Return (V_{UNLEV})	$\Delta TEV \cdot \overline{FO}$
Cash Flow Generation (V_{CF})	$-\Delta ND \cdot \overline{FO}$
EBITDA Growth (V_{EG})	$\Delta E \cdot \overline{M} \cdot \overline{FO}$
Multiple Expansion (V_{ME})	$\Delta M \cdot \overline{E} \cdot \overline{FO}$
Revenue Growth (V_{RG})	$\Delta R \cdot \overline{EM} \cdot \overline{M} \cdot \overline{FO}$
EBITDA Margin Expansion (V_{EME})	$\Delta EM \cdot \overline{R} \cdot \overline{M} \cdot \overline{FO}$
Gross Margin Expansion (V_{GME})	$\Delta GM \cdot \overline{R} \cdot \overline{M} \cdot \overline{FO}$
Operating Margin Reduction (V_{OMR})	$-\Delta OM \cdot \overline{R} \cdot \overline{M} \cdot \overline{FO}$
Addressable Market Growth (V_{AMG})	$\frac{\beta_R}{2}\left(R_1 + \left(\frac{R_2}{1+\beta_R}\right)\right)\overline{EM} \cdot \overline{M} \cdot \overline{FO}$
Market Share Growth (V_{MSG})	$\frac{\alpha_R}{2}\left(R_1 + \left(\frac{R_2}{1+\alpha_R}\right)\right)\overline{EM} \cdot \overline{M} \cdot \overline{FO}$
Market Multiple Expansion (V_{MME})	$\Delta MM \cdot \overline{\mu} \cdot \overline{E} \cdot \overline{FO}$
Intrinsic Multiple Expansion (V_{IME})	$\Delta \mu \cdot \overline{MM} \cdot \overline{E} \cdot \overline{FO}$
Net Return (V_{NET})	$FEqV_2 - FEqV_1 \cdot (1 + \%F)$ $-\%C \cdot MAX[0, FEqV_2 - FEqV_1 \cdot (1 + \%F)]$
Fees & Expenses (V_{FEES})	$\%F \cdot FEqV_1$
Carried Interest (V_{CARRY})	$\%C \cdot MAX[0, FEqV_2 - FEqV_1 \cdot (1 + \%F)]$

Relative (%) value creation (υ) is absolute ($) value creation ($V$) divided by the change in the Fund investment's equity value ($\Delta FEqV$).

Table 2.6 Conventional model relative (%) returns and value drivers

Gross Return (υ_{GROSS})	100%
Fund Ownership Impact (υ_{FO})	$\Delta FO \cdot \overline{TEqV}/\Delta FEqV$
Isomeric Return (υ_{ISO})	$\Delta TEqV \cdot \overline{FO}/\Delta FEqV$
Unlevered Return (υ_{UNLEV})	$\Delta TEV \cdot \overline{FO}/\Delta FEqV$
Cash Flow Generation (υ_{CF})	$-\Delta ND \cdot \overline{FO}/\Delta FEqV$
EBITDA Growth (υ_{EG})	$\Delta E \cdot \overline{M} \cdot \overline{FO}/\Delta FEqV$
Multiple Expansion (υ_{ME})	$\Delta M \cdot \overline{E} \cdot \overline{FO}/\Delta FEqV$
Revenue Growth (υ_{RG})	$\Delta R \cdot \overline{EM} \cdot \overline{M} \cdot \overline{FO}/\Delta FEqV$
EBITDA Margin Expansion (υ_{EME})	$\Delta EM \cdot \overline{R} \cdot \overline{M} \cdot \overline{FO}/\Delta FEqV$
Gross Margin Expansion (υ_{GME})	$\Delta GM \cdot \overline{R} \cdot \overline{M} \cdot \overline{FO}/\Delta FEqV$
Operating Margin Reduction (υ_{OMR})	$-\Delta OM \cdot \overline{R} \cdot \overline{M} \cdot \overline{FO}/\Delta FEqV$
Addressable Market Growth (υ_{AMG})	$\frac{\beta_R}{2}\left(R_1 + \left(\frac{R_2}{1+\beta_R}\right)\right)\frac{\overline{EM} \cdot \overline{M} \cdot \overline{FO}}{\Delta FEqV}$
Market Share Growth (υ_{MSG})	$\frac{\alpha_R}{2}\left(R_1 + \left(\frac{R_2}{1+\alpha_R}\right)\right)\frac{\overline{EM} \cdot \overline{M} \cdot \overline{FO}}{\Delta FEqV}$
Market Multiple Expansion (υ_{MME})	$\Delta MM \cdot \overline{\mu} \cdot \overline{E} \cdot \overline{FO}/\Delta FEqV$
Intrinsic Multiple Expansion (υ_{IME})	$\Delta \mu \cdot \overline{MM} \cdot \overline{E} \cdot \overline{FO}/\Delta FEqV$
Net Return (υ_{NET})	$\frac{\left(\begin{array}{c} FEqV_2 - FEqV_1 \cdot (1+\%F) \\ -\%C \cdot MAX[0, FEqV_2 - FEqV_1 \cdot (1+\%F)] \end{array}\right)}{\Delta FEqV}$
Fees & Expenses (υ_{FEES})	$\%F \cdot FEqV_1/\Delta FEqV$
Carried Interest (υ_{CARRY})	$\frac{\%C \cdot MAX[0, FEqV_2 - FEqV_1 \cdot (1+\%F)]}{\Delta FEqV}$

Equity return multiples or multipliers (χ) are the Fund investment's Gross Multiple of Invested Capital (MOIC) raised to the power of each return or value driver's relative (%) value creation (υ).

Table 2.7 Conventional model equity return multipliers

Gross Return ($MOIC$, χ_{GROSS})	$FEqV_2/FEqV_1$
Fund Ownership Impact (χ_{FO})	$MOIC^{\frac{\Delta FO\cdot\overline{TEqV}}{\Delta FEqV}}$
Isomeric Return (χ_{ISO})	$MOIC^{\frac{\Delta TEqV\cdot\overline{FO}}{\Delta FEqV}}$
Unlevered Return (χ_{UNLEV})	$MOIC^{\frac{\Delta TEV\cdot\overline{FO}}{\Delta FEqV}}$
Cash Flow Generation (χ_{CF})	$MOIC^{\frac{-\Delta ND\cdot\overline{FO}}{\Delta FEqV}}$
EBITDA Growth (χ_{EG})	$MOIC^{\frac{\Delta E\cdot\overline{M}\cdot\overline{FO}}{\Delta FEqV}}$
Multiple Expansion (χ_{ME})	$MOIC^{\frac{\Delta M\cdot\overline{E}\cdot\overline{FO}}{\Delta FEqV}}$
Revenue Growth (χ_{RG})	$MOIC^{\frac{\Delta R\cdot\overline{EM}\cdot\overline{M}\cdot\overline{FO}}{\Delta FEqV}}$
EBITDA Margin Expansion (χ_{EME})	$MOIC^{\frac{\Delta EM\cdot\overline{R}\cdot\overline{M}\cdot\overline{FO}}{\Delta FEqV}}$
Gross Margin Expansion (χ_{GME})	$MOIC^{\frac{\Delta GM\cdot\overline{R}\cdot\overline{M}\cdot\overline{FO}}{\Delta FEqV}}$
Operating Margin Reduction (χ_{OMR})	$MOIC^{\frac{-\Delta OM\cdot\overline{R}\cdot\overline{M}\cdot\overline{FO}}{\Delta FEqV}}$
Addressable Market Growth (χ_{AMG})	$MOIC^{\frac{\beta_R}{2}\left(R_1+\left(\frac{R_2}{1+\beta_R}\right)\right)\frac{\overline{EM}\cdot\overline{M}\cdot\overline{FO}}{\Delta FEqV}}$
Market Share Growth (χ_{MSG})	$MOIC^{\frac{\alpha_R}{2}\left(R_1+\left(\frac{R_2}{1+\alpha_R}\right)\right)\frac{\overline{EM}\cdot\overline{M}\cdot\overline{FO}}{\Delta FEqV}}$
Market Multiple Expansion (χ_{MME})	$MOIC^{\frac{\Delta MM\cdot\overline{\mu}\cdot\overline{E}\cdot\overline{FO}}{\Delta FEqV}}$
Intrinsic Multiple Expansion (χ_{IME})	$MOIC^{\frac{\Delta\mu\cdot\overline{MM}\cdot\overline{E}\cdot\overline{FO}}{\Delta FEqV}}$
Net Return ($MOIC_{NET}$, χ_{NET})	$MOIC^{\frac{\left(\begin{array}{c}FEqV_2-FEqV_1\cdot(1+\%F)\\-\%C\cdot MAX\left[0,FEqV_2-FEqV_1\cdot(1+\%F)\right]\end{array}\right)}{\Delta FEqV}}$
Fees & Expenses (χ_{FEES})	$MOIC^{\frac{\%F\cdot FEqV_1}{\Delta FEqV}}$
Carried Interest (χ_{CARRY})	$MOIC^{\frac{\%C\cdot MAX[0,FEqV_2-FEqV_1\cdot(1+\%F)]}{\Delta FEqV}}$

Value creation IRRs are each return or value driver's equity return multiple or multiplier (χ) raised to the power of the reciprocal hold (Δt) in years, minus 100%.

Table 2.8 Conventional model value creation IRRs

Gross Return (IRR, IRR_{GROSS})	$MOIC^{1/\Delta t}-1$
Fund Ownership Impact (IRR_{FO})	$MOIC^{\frac{\Delta FO\cdot\overline{TEqV}}{\Delta FEqV\cdot\Delta t}}-1$
Isomeric Return (IRR_{ISO})	$MOIC^{\frac{\Delta TEqV\cdot\overline{FO}}{\Delta FEqV\cdot\Delta t}}-1$
Unlevered Return (IRR_{UNLEV})	$MOIC^{\frac{\Delta TEV\cdot\overline{FO}}{\Delta FEqV\cdot\Delta t}}-1$
Cash Flow Generation (IRR_{CF})	$MOIC^{\frac{-\Delta ND\cdot\overline{FO}}{\Delta FEqV\cdot\Delta t}}-1$
EBITDA Growth (IRR_{EG})	$MOIC^{\frac{\Delta E\cdot\overline{M}\cdot\overline{FO}}{\Delta FEqV\cdot\Delta t}}-1$
Multiple Expansion (IRR_{ME})	$MOIC^{\frac{\Delta M\cdot\overline{E}\cdot\overline{FO}}{\Delta FEqV\cdot\Delta t}}-1$
Revenue Growth (IRR_{RG})	$MOIC^{\frac{\Delta R\cdot\overline{EM}\cdot\overline{M}\cdot\overline{FO}}{\Delta FEqV\cdot\Delta t}}-1$
EBITDA Margin Expansion (IRR_{EME})	$MOIC^{\frac{\Delta EM\cdot\overline{R}\cdot\overline{M}\cdot\overline{FO}}{\Delta FEqV\cdot\Delta t}}-1$
Gross Margin Expansion (IRR_{GME})	$MOIC^{\frac{\Delta GM\cdot R\cdot\overline{M}\cdot\overline{FO}}{\Delta FEqV\cdot\Delta t}}-1$
Operating Margin Reduction (IRR_{OMR})	$MOIC^{\frac{-\Delta OM\cdot\overline{R}\cdot\overline{M}\cdot\overline{FO}}{\Delta FEqV\cdot\Delta t}}-1$
Addressable Market Growth (IRR_{AMG})	$MOIC^{\frac{\beta_R}{2}\left(R_1+\left(\frac{R_2}{1+\beta_R}\right)\right)\frac{\overline{EM}\cdot\overline{M}\cdot\overline{FO}}{\Delta FEqV\cdot\Delta t}}-1$
Market Share Growth (IRR_{MSG})	$MOIC^{\frac{\alpha_R}{2}\left(R_1+\left(\frac{R_2}{1+\alpha_R}\right)\right)\frac{\overline{EM}\cdot\overline{M}\cdot\overline{FO}}{\Delta FEqV\cdot\Delta t}}-1$
Market Multiple Expansion (IRR_{MME})	$MOIC^{\frac{\Delta MM\cdot\overline{\mu}\cdot\overline{E}\cdot\overline{FO}}{\Delta FEqV\cdot\Delta t}}-1$
Intrinsic Multiple Expansion (IRR_{IME})	$MOIC^{\frac{\Delta\mu\cdot\overline{MM}\cdot\overline{E}\cdot\overline{FO}}{\Delta FEqV\cdot\Delta t}}-1$
Net Return (IRR_{NET})	$MOIC^{\frac{\left(\begin{array}{c}FEqV_2-FEqV_1\cdot(1+\%F)\\-\%C\cdot MAX\left[0,FEqV_2-FEqV_1\cdot(1+\%F)\right]\end{array}\right)}{\Delta FEqV\cdot\Delta t}}-1$
Fees & Expenses (IRR_{FEES})	$MOIC^{\frac{\%F\cdot FEqV_1}{\Delta FEqV\cdot\Delta t}}-1$
Carried Interest (IRR_{CARRY})	$MOIC^{\frac{\%C\cdot MAX[0,FEqV_2-FEqV_1\cdot(1+\%F)]}{\Delta FEqV\cdot\Delta t}}-1$

2.3. Derivative Model of Value Creation

Derivative Model returns and value drivers are calculated first as absolute ($) value creation ($V$), and then converted into relative (%) value creation (v), equity return multiples or multipliers (χ), and value creation IRRs.

Table 2.9 Derivative model absolute ($) returns and value drivers

Gross Return (V_{GROSS})	$\Delta FEqV$
Fund Ownership Impact (V_{FO})	$\Delta FO \cdot \overline{TEqV}$
Isomeric Return (V_{ISO})	$\Delta TEqV \cdot \overline{FO}$
Unlevered Return (V_{UNLEV})	$\Delta TEV \cdot \overline{ER} \cdot \overline{FO}$
Leverage Effect (V_{LEV})	$\Delta ER \cdot \overline{TEV} \cdot \overline{FO}$
EBITDA Growth (V_{EG})	$\Delta E \cdot \overline{M} \cdot \overline{ER} \cdot \overline{FO}$
Multiple Expansion (V_{ME})	$\Delta M \cdot \overline{E} \cdot \overline{ER} \cdot \overline{FO}$
Revenue Growth (V_{RG})	$\Delta R \cdot \overline{EM} \cdot \overline{M} \cdot \overline{ER} \cdot \overline{FO}$
EBITDA Margin Expansion (V_{EME})	$\Delta EM \cdot \overline{R} \cdot \overline{M} \cdot \overline{ER} \cdot \overline{FO}$
Gross Margin Expansion (V_{GME})	$\Delta GM \cdot \overline{R} \cdot \overline{M} \cdot \overline{ER} \cdot \overline{FO}$
Operating Margin Reduction (V_{OMR})	$-\Delta OM \cdot \overline{R} \cdot \overline{M} \cdot \overline{ER} \cdot \overline{FO}$
Addressable Market Growth (V_{AMG})	$\frac{\beta_R}{2}\left(R_1 + \left(\frac{R_2}{1+\beta_R}\right)\right)\overline{EM} \cdot \overline{M} \cdot \overline{ER} \cdot \overline{FO}$
Market Share Growth (V_{MSG})	$\frac{\alpha_R}{2}\left(R_1 + \left(\frac{R_2}{1+\alpha_R}\right)\right)\overline{EM} \cdot \overline{M} \cdot \overline{ER} \cdot \overline{FO}$
Market Multiple Expansion (V_{MME})	$\Delta MM \cdot \overline{\mu} \cdot \overline{E} \cdot \overline{ER} \cdot \overline{FO}$
Intrinsic Multiple Expansion (V_{IME})	$\Delta \mu \cdot \overline{MM} \cdot \overline{E} \cdot \overline{ER} \cdot \overline{FO}$
Net Return (V_{NET})	$FEqV_2 - FEqV_1 \cdot (1 + \%F)$ $-\%C \cdot MAX[0, FEqV_2 - FEqV_1 \cdot (1 + \%F)]$
Fees & Expenses (V_{FEES})	$\%F \cdot FEqV_1$
Carried Interest (V_{CARRY})	$\%C \cdot MAX[0, FEqV_2 - FEqV_1 \cdot (1 + \%F)]$

Relative (%) value creation (υ) is absolute ($) value creation ($V$) divided by the change in the Fund investment's equity value ($\Delta FEqV$).

Table 2.10 Derivative model relative (%) returns and value drivers

Gross Return (υ_{GROSS})	100%
Fund Ownership Impact (υ_{FO})	$\Delta FO \cdot \overline{TEqV} / \Delta FEqV$
Isomeric Return (υ_{ISO})	$\Delta TEqV \cdot \overline{FO} / \Delta FEqV$
Unlevered Return (υ_{UNLEV})	$\Delta TEV \cdot \overline{ER} \cdot \overline{FO} / \Delta FEqV$
Leverage Effect (υ_{LEV})	$\Delta ER \cdot \overline{TEV} \cdot \overline{FO} / \Delta FEqV$
EBITDA Growth (υ_{EG})	$\Delta E \cdot \overline{M} \cdot \overline{ER} \cdot \overline{FO} / \Delta FEqV$
Multiple Expansion (υ_{ME})	$\Delta M \cdot \overline{E} \cdot \overline{ER} \cdot \overline{FO} / \Delta FEqV$
Revenue Growth (υ_{RG})	$\Delta R \cdot \overline{EM} \cdot \overline{M} \cdot \overline{ER} \cdot \overline{FO} / \Delta FEqV$
EBITDA Margin Expansion (υ_{EME})	$\Delta EM \cdot \overline{R} \cdot \overline{M} \cdot \overline{ER} \cdot \overline{FO} / \Delta FEqV$
Gross Margin Expansion (υ_{GME})	$\Delta GM \cdot \overline{R} \cdot \overline{M} \cdot \overline{ER} \cdot \overline{FO} / \Delta FEqV$
Operating Margin Reduction (υ_{OMR})	$-\Delta OM \cdot \overline{R} \cdot \overline{M} \cdot \overline{ER} \cdot \overline{FO} / \Delta FEqV$
Addressable Market Growth (υ_{AMG})	$\frac{\beta_R}{2}\left(R_1 + \left(\frac{R_2}{1+\beta_R}\right)\right)\frac{\overline{EM} \cdot \overline{M} \cdot \overline{ER} \cdot \overline{FO}}{\Delta FEqV}$
Market Share Growth (υ_{MSG})	$\frac{\alpha_R}{2}\left(R_1 + \left(\frac{R_2}{1+\alpha_R}\right)\right)\frac{\overline{EM} \cdot \overline{M} \cdot \overline{ER} \cdot \overline{FO}}{\Delta FEqV}$
Market Multiple Expansion (υ_{MME})	$\Delta MM \cdot \overline{\mu} \cdot \overline{E} \cdot \overline{ER} \cdot \overline{FO} / \Delta FEqV$
Intrinsic Multiple Expansion (υ_{IME})	$\Delta \mu \cdot \overline{MM} \cdot \overline{E} \cdot \overline{ER} \cdot \overline{FO} / \Delta FEqV$
Net Return (υ_{NET})	$\frac{\left(\begin{array}{c} FEqV_2 - FEqV_1 \cdot (1+\%F) \\ -\%C \cdot MAX[0, FEqV_2 - FEqV_1 \cdot (1+\%F)] \end{array}\right)}{\Delta FEqV}$
Fees & Expenses (υ_{FEES})	$\frac{\%F \cdot FEqV_1}{\Delta FEqV}$
Carried Interest (υ_{CARRY})	$\frac{\%C \cdot MAX[0, FEqV_2 - FEqV_1 \cdot (1+\%F)]}{\Delta FEqV}$

Equity return multiples or multipliers (χ) are the Fund investment's Gross Multiple of Invested Capital (MOIC) raised to the power of each return or value driver's relative (%) value creation (υ).

Table 2.11 Derivative model equity return multipliers

Gross Return ($MOIC$, χ_{GROSS})	$FEqV_2/FEqV_1$
Fund Ownership Impact (χ_{FO})	$MOIC^{\frac{\Delta FO\cdot\overline{TEqV}}{\Delta FEqV}}$
Isomeric Return (χ_{ISO})	$MOIC^{\frac{\Delta TEqV\cdot\overline{FO}}{\Delta FEqV}}$
Unlevered Return (χ_{UNLEV})	$MOIC^{\frac{\Delta TEV\cdot\overline{ER}\cdot\overline{FO}}{\Delta FEqV}}$
Leverage Effect (χ_{LEV})	$MOIC^{\frac{\Delta ER\cdot\overline{TEV}\cdot\overline{FO}}{\Delta FEqV}}$
EBITDA Growth (χ_{EG})	$MOIC^{\frac{\Delta E\cdot\overline{M}\cdot\overline{ER}\cdot\overline{FO}}{\Delta FEqV}}$
Multiple Expansion (χ_{ME})	$MOIC^{\frac{\Delta M\cdot\overline{E}\cdot\overline{ER}\cdot\overline{FO}}{\Delta FEqV}}$
Revenue Growth (χ_{RG})	$MOIC^{\frac{\Delta R\cdot\overline{EM}\cdot\overline{M}\cdot\overline{ER}\cdot\overline{FO}}{\Delta FEqV}}$
EBITDA Margin Expansion (χ_{EME})	$MOIC^{\frac{\Delta EM\cdot\overline{R}\cdot\overline{M}\cdot\overline{ER}\cdot\overline{FO}}{\Delta FEqV}}$
Gross Margin Expansion (χ_{GME})	$MOIC^{\frac{\Delta GM\cdot\overline{R}\cdot\overline{M}\cdot\overline{ER}\cdot\overline{FO}}{\Delta FEqV}}$
Operating Margin Reduction (χ_{OMR})	$MOIC^{\frac{-\Delta OM\cdot\overline{R}\cdot\overline{M}\cdot\overline{ER}\cdot\overline{FO}}{\Delta FEqV}}$
Addressable Market Growth (χ_{AMG})	$MOIC^{\frac{\beta_R}{2}\left(R_1+\left(\frac{R_2}{1+\beta_R}\right)\right)\frac{\overline{EM}\cdot\overline{M}\cdot\overline{ER}\cdot\overline{FO}}{\Delta FEqV}}$
Market Share Growth (χ_{MSG})	$MOIC^{\frac{\alpha_R}{2}\left(R_1+\left(\frac{R_2}{1+\alpha_R}\right)\right)\frac{\overline{EM}\cdot\overline{M}\cdot\overline{ER}\cdot\overline{FO}}{\Delta FEqV}}$
Market Multiple Expansion (χ_{MME})	$MOIC^{\frac{\Delta MM\cdot\overline{\mu}\cdot\overline{E}\cdot\overline{ER}\cdot\overline{FO}}{\Delta FEqV}}$
Intrinsic Multiple Expansion (χ_{IME})	$MOIC^{\frac{\Delta\mu\cdot\overline{MM}\cdot\overline{E}\cdot\overline{ER}\cdot\overline{FO}}{\Delta FEqV}}$
Net Return ($MOIC_{NET}$, χ_{NET})	$MOIC^{\frac{\left(\begin{array}{c}FEqV_2-FEqV_1\cdot(1+\%F)\\-\%C\cdot MAX\left[0,FEqV_2-FEqV_1\cdot(1+\%F)\right]\end{array}\right)}{\Delta FEqV}}$
Fees & Expenses (χ_{FEES})	$MOIC^{\frac{\%F\cdot FEqV_1}{\Delta FEqV}}$
Carried Interest (χ_{CARRY})	$MOIC^{\frac{\%C\cdot MAX[0,FEqV_2-FEqV_1\cdot(1+\%F)]}{\Delta FEqV}}$

Value creation IRRs are each return or value driver's equity return multiple or multiplier (χ) raised to the power of the reciprocal hold (Δt) in years, minus 100%.

Table 2.12 Derivative model value creation IRRs

Gross Return (IRR, IRR_{GROSS})	$MOIC^{1/\Delta t} - 1$
Fund Ownership Impact (IRR_{FO})	$MOIC^{\frac{\Delta FO \cdot \overline{TEqV}}{\Delta FEqV \cdot \Delta t}} - 1$
Isomeric Return (IRR_{ISO})	$MOIC^{\frac{\Delta TEqV \cdot \overline{FO}}{\Delta FEqV \cdot \Delta t}} - 1$
Unlevered Return (IRR_{UNLEV})	$MOIC^{\frac{\Delta TEV \cdot \overline{ER} \cdot \overline{FO}}{\Delta FEqV \cdot \Delta t}} - 1$
Leverage Effect (IRR_{LEV})	$MOIC^{\frac{\Delta ER \cdot \overline{TEV} \cdot \overline{FO}}{\Delta FEqV \cdot \Delta t}} - 1$
EBITDA Growth (IRR_{EG})	$MOIC^{\frac{\Delta E \cdot \overline{M} \cdot \overline{ER} \cdot \overline{FO}}{\Delta FEqV \cdot \Delta t}} - 1$
Multiple Expansion (IRR_{ME})	$MOIC^{\frac{\Delta M \cdot \overline{E} \cdot \overline{ER} \cdot \overline{FO}}{\Delta FEqV \cdot \Delta t}} - 1$
Revenue Growth (IRR_{RG})	$MOIC^{\frac{\Delta R \cdot \overline{EM} \cdot \overline{M} \cdot \overline{ER} \cdot \overline{FO}}{\Delta FEqV \cdot \Delta t}} - 1$
EBITDA Margin Expansion (IRR_{EME})	$MOIC^{\frac{\Delta EM \cdot \overline{R} \cdot \overline{M} \cdot \overline{ER} \cdot \overline{FO}}{\Delta FEqV \cdot \Delta t}} - 1$
Gross Margin Expansion (IRR_{GME})	$MOIC^{\frac{\Delta GM \cdot \overline{R} \cdot \overline{M} \cdot \overline{ER} \cdot \overline{FO}}{\Delta FEqV \cdot \Delta t}} - 1$
Operating Margin Reduction (IRR_{OMR})	$MOIC^{\frac{-\Delta OM \cdot \overline{R} \cdot \overline{M} \cdot \overline{ER} \cdot \overline{FO}}{\Delta FEqV \cdot \Delta t}} - 1$
Addressable Market Growth (IRR_{AMG})	$MOIC^{\frac{\beta_R}{2}\left(R_1+\left(\frac{R_2}{1+\beta_R}\right)\right)\frac{\overline{EM} \cdot \overline{M} \cdot \overline{ER} \cdot \overline{FO}}{\Delta FEqV \cdot \Delta t}} - 1$
Market Share Growth (IRR_{MSG})	$MOIC^{\frac{\alpha_R}{2}\left(R_1+\left(\frac{R_2}{1+\alpha_R}\right)\right)\frac{\overline{EM} \cdot \overline{M} \cdot \overline{ER} \cdot \overline{FO}}{\Delta FEqV \cdot \Delta t}} - 1$
Market Multiple Expansion (IRR_{MME})	$MOIC^{\frac{\Delta MM \cdot \overline{\mu} \cdot \overline{E} \cdot \overline{ER} \cdot \overline{FO}}{\Delta FEqV \cdot \Delta t}} - 1$
Intrinsic Multiple Expansion (IRR_{IME})	$MOIC^{\frac{\Delta \mu \cdot \overline{MM} \cdot \overline{E} \cdot \overline{ER} \cdot \overline{FO}}{\Delta FEqV \cdot \Delta t}} - 1$
Net Return (IRR_{NET})	$MOIC^{\frac{\left(\begin{array}{c} FEqV_2 - FEqV_1 \cdot (1+\%F) \\ -\%C \cdot MAX\left[0, FEqV_2 - FEqV_1 \cdot (1+\%F)\right] \end{array}\right)}{\Delta FEqV \cdot \Delta t}} - 1$
Fees & Expenses (IRR_{FEES})	$MOIC^{\frac{\%F \cdot FEqV_1}{\Delta FEqV \cdot \Delta t}} - 1$
Carried Interest (IRR_{CARRY})	$MOIC^{\frac{\%C \cdot MAX[0, FEqV_2 - FEqV_1 \cdot (1+\%F)]}{\Delta FEqV \cdot \Delta t}} - 1$

2.4. Logarithmic Model of Value Creation

Logarithmic Model returns and value drivers are calculated first as equity return multiples or multipliers (χ), and then converted into either relative (%) value creation (v) or value creation IRRs. Absolute ($) value creation ($V$) is derived from relative (%) value creation.

Table 2.13 Logarithmic model equity return multipliers

Gross Return ($MOIC$, χ_{GROSS})	$FEqV_2/FEqV_1$
Fund Ownership Impact (χ_{FO})	FO_2/FO_1
Isomeric Return (χ_{ISO})	$TEqV_2/TEqV_1$
Unlevered Return (χ_{UNLEV})	TEV_2/TEV_1
Leverage Effect (χ_{LEV})	ER_2/ER_1
EBITDA Growth (χ_{EG})	E_2/E_1
Multiple Expansion (χ_{ME})	M_2/M_1
Revenue Growth (χ_{RG})	R_2/R_1
EBITDA Margin Expansion (χ_{EME})	EM_2/EM_1
Gross Margin Expansion (χ_{GME})	$e^{\Delta GM/\overline{EM}}$
Operating Margin Reduction (χ_{OMR})	$e^{-\Delta OM/\overline{EM}}$
Addressable Market Growth (χ_{AMG})	$1+\beta_R$
Market Share Growth (χ_{MSG})	$1+\alpha_R$
Market Multiple Expansion (χ_{MME})	MM_2/MM_1
Intrinsic Multiple Expansion (χ_{IME})	μ_2/μ_1
Net Return ($MOIC_{NET}$, χ_{NET})	$\dfrac{MOIC \cdot MIN\left[1{,}1-\%C+\dfrac{\%C\cdot(1+\%F)}{MOIC}\right]}{(1+\%F)}$
Fees & Expenses (χ_{FEES})	$1+\%F$
Carried Interest (χ_{CARRY})	$MAX\left[1,\dfrac{1}{1-\%C+\dfrac{\%C\cdot(1+\%F)}{MOIC}}\right]$

Value creation IRRs are each return or value driver's equity return multiple or multiplier (χ) raised to the power of the reciprocal hold (Δt) in years, minus 100%.

Table 2.14 Logarithmic model value creation IRRs

Gross Return (IRR, IRR_{GROSS})	$MOIC^{1/\Delta t} - 1$
Fund Ownership Impact (IRR_{FO})	$(FO_2/FO_1)^{1/\Delta t} - 1$
Isomeric Return (IRR_{ISO})	$(TEqV_2/TEqV_1)^{1/\Delta t} - 1$
Unlevered Return (IRR_{UNLEV})	$(TEV_2/TEV_1)^{1/\Delta t} - 1$
Leverage Effect (IRR_{LEV})	$(ER_2/ER_1)^{1/\Delta t} - 1$
EBITDA Growth (IRR_{EG})	$(E_2/E_1)^{1/\Delta t} - 1$
Multiple Expansion (IRR_{ME})	$(M_2/M_1)^{1/\Delta t} - 1$
Revenue Growth (IRR_{RG})	$(R_2/R_1)^{1/\Delta t} - 1$
EBITDA Margin Expansion (IRR_{EME})	$(EM_2/EM_1)^{1/\Delta t} - 1$
Gross Margin Expansion (IRR_{GME})	$e^{\Delta GM/\overline{EM}\cdot\Delta t} - 1$
Operating Margin Reduction (IRR_{OMR})	$e^{-\Delta OM/\overline{EM}\cdot\Delta t} - 1$
Addressable Market Growth (IRR_{AMG})	$(1 + \beta_R)^{1/\Delta t} - 1$
Market Share Growth (IRR_{MSG})	$(1 + \alpha_R)^{1/\Delta t} - 1$
Market Multiple Expansion (IRR_{MME})	$(MM_2/MM_1)^{1/\Delta t} - 1$
Intrinsic Multiple Expansion (IRR_{IME})	$(\mu_2/\mu_1)^{1/\Delta t} - 1$
Net Return (IRR_{NET})	$\left\{\dfrac{MOIC\cdot MIN\left[1,1-\%C+\dfrac{\%C\cdot(1+\%F)}{MOIC}\right]}{(1+\%F)}\right\}^{1/\Delta t} - 1$
Fees & Expenses (IRR_{FEES})	$(1 + \%F)^{1/\Delta t} - 1$
Carried Interest (IRR_{CARRY})	$MAX\left[1,\dfrac{1}{1-\%C+\dfrac{\%C\cdot(1+\%F)}{MOIC}}\right]^{1/\Delta t} - 1$

Relative (%) value creation (υ) is the natural logarithm of the return or value driver's equity return multiple or multiplier (χ) divided by the natural logarithm of the Fund investment's Gross Multiple of Invested Capital (MOIC).

Table 2.15 Logarithmic model relative (%) returns and value drivers

Gross Return (υ_{GROSS})	100%
Fund Ownership Impact (υ_{FO})	$\ln(FO_2/FO_1)/\ln(MOIC)$
Isomeric Return (υ_{ISO})	$\ln(TEqV_2/TEqV_1)/\ln(MOIC)$
Unlevered Return (υ_{UNLEV})	$\ln(TEV_2/TEV_1)/\ln(MOIC)$
Leverage Effect (υ_{LEV})	$\ln(ER_2/ER_1)/\ln(MOIC)$
EBITDA Growth (υ_{EG})	$\ln(E_2/E_1)/\ln(MOIC)$
Multiple Expansion (υ_{ME})	$\ln(M_2/M_1)/\ln(MOIC)$
Revenue Growth (υ_{RG})	$\ln(R_2/R_1)/\ln(MOIC)$
EBITDA Margin Expansion (υ_{EME})	$\ln(EM_2/EM_1)/\ln(MOIC)$
Gross Margin Expansion (υ_{GME})	$(\Delta GM/\overline{\mathrm{EM}})/\ln(MOIC)$
Operating Margin Reduction (υ_{OMR})	$(-\Delta OM/\overline{\mathrm{EM}})/\ln(MOIC)$
Addressable Market Growth (υ_{AMG})	$\ln(1+\beta_R)/\ln(MOIC)$
Market Share Growth (υ_{MSG})	$\ln(1+\alpha_R)/\ln(MOIC)$
Market Multiple Expansion (υ_{MME})	$\ln(MM_2/MM_1)/\ln(MOIC)$
Intrinsic Multiple Expansion (υ_{IME})	$\ln(\mu_2/\mu_1)/\ln(MOIC)$
Net Return (υ_{NET})	$\dfrac{\ln\left(\dfrac{MOIC \cdot MIN\left[1, 1-\%C+\dfrac{\%C\cdot(1+\%F)}{MOIC}\right]}{(1+\%F)}\right)}{\ln(MOIC)}$
Fees & Expenses (υ_{FEES})	$\ln(1+\%F)/\ln(MOIC)$
Carried Interest (υ_{CARRY})	$\dfrac{\ln\left(MAX\left[1, \dfrac{1}{1-\%C+\dfrac{\%C\cdot(1+\%F)}{MOIC}}\right]\right)}{\ln(MOIC)}$

Absolute ($) value creation ($V$) is relative (%) value creation (v) multiplied by the change in the Fund investment's equity value ($\Delta FEqV$).

Table 2.16 Logarithmic model absolute ($) returns and value drivers

Gross Return (V_{GROSS})	$\Delta FEqV$
Fund Ownership Impact (V_{FO})	$\Delta FEqV \cdot \ln(FO_2/FO_1)/\ln(MOIC)$
Isomeric Return (V_{ISO})	$\Delta FEqV \cdot \ln(TEqV_2/TEqV_1)/\ln(MOIC)$
Unlevered Return (V_{UNLEV})	$\Delta FEqV \cdot \ln(TEV_2/TEV_1)/\ln(MOIC)$
Leverage Effect (V_{LEV})	$\Delta FEqV \cdot \ln(ER_2/ER_1)/\ln(MOIC)$
EBITDA Growth (V_{EG})	$\Delta FEqV \cdot \ln(E_2/E_1)/\ln(MOIC)$
Multiple Expansion (V_{ME})	$\Delta FEqV \cdot \ln(M_2/M_1)/\ln(MOIC)$
Revenue Growth (V_{RG})	$\Delta FEqV \cdot \ln(R_2/R_1)/\ln(MOIC)$
EBITDA Margin Expansion (V_{EME})	$\Delta FEqV \cdot \ln(EM_2/EM_1)/\ln(MOIC)$
Gross Margin Expansion (V_{GME})	$\Delta FEqV \cdot (\Delta GM/\overline{\text{EM}})/\ln(MOIC)$
Operating Margin Reduction (V_{OMR})	$\Delta FEqV \cdot (-\Delta OM/\overline{\text{EM}})/\ln(MOIC)$
Addressable Market Growth (V_{AMG})	$\Delta FEqV \cdot \ln(1+\beta_R)/\ln(MOIC)$
Market Share Growth (V_{MSG})	$\Delta FEqV \cdot \ln(1+\alpha_R)/\ln(MOIC)$
Market Multiple Expansion (V_{MME})	$\Delta FEqV \cdot \ln(MM_2/MM_1)/\ln(MOIC)$
Intrinsic Multiple Expansion (V_{IME})	$\Delta FEqV \cdot \ln(\mu_2/\mu_1)/\ln(MOIC)$
Net Return (V_{NET})	$\dfrac{\Delta FEqV}{\ln(MOIC)} \cdot \ln\left(\dfrac{MOIC \cdot MIN\left[1, 1-\%C+\dfrac{\%C\cdot(1+\%F)}{MOIC}\right]}{(1+\%F)}\right)$
Fees & Expenses (V_{FEES})	$\dfrac{\Delta FEqV}{\ln(MOIC)} \cdot \ln(1+\%F)$
Carried Interest (V_{CARRY})	$\dfrac{\Delta FEqV}{\ln(MOIC)} \cdot \ln\left(MAX\left[1, \dfrac{1}{1-\%C+\dfrac{\%C\cdot(1+\%F)}{MOIC}}\right]\right)$

3. Value Creation Principles and Best Practices

This chapter describes a few of the principles and best practices required to develop value creation models that are mathematically rigorous, rely on accessible data with which GPs and LPs are familiar, and provide sensible and non-volatile results over the largest range of company, industry, and market conditions.

3.1. Value Creation is Objective

Value creation is equity value change. It is an objective direct measurement. *Private Equity Value Creation Analysis* assumes that equity valuations are real. They are measurable, like length. In physics, a change in position from point X_1 to X_2 is "displacement" (ΔX). Similarly, a change in private company equity value is "value creation" ($\Delta TEqV$ or $\Delta FEqV$).

Assume that between times t_1 and t_2 company total equity value ($TEqV$) grows from $TEqV_1$ to $TEqV_2$, while the Fund's investment, Fund equity value ($FEqV$), grows from $FEqV_1$ to $FEqV_2$. These equity value changes represent value creation and are expressed:

$$\Delta TEqV = TEqV_2 - TEqV_1 = V_{EQUITY}$$

$$\Delta FEqV = FEqV_2 - FEqV_1 = V_{GROSS}$$

Value creation is also commonly expressed as a Multiple of Invested Capital (MOIC), where:

$$MOIC_{EQUITY} = TEqV_2/TEqV_1 = X_{EQUITY}$$

$$MOIC_{GROSS} = FEqV_2/FEqV_1 = X_{GROSS}$$

The subscripts distinguish value creation for the entire shareholder group (*EQUITY*) from value creation for a specific shareholder or Fund (*GROSS*). The *FEqV*-terms correspond to the most common Fund-level return measurements usually found in Fund reporting and GP marketing materials (e.g., Gross Invested Capital, Gross MOIC, Gross IRR, etc.).

3.2. Value Creation Analysis is Subjective

Value creation analysis is an interpretation of why equity values change. It is a subjective indirect inference. The following chapters describe about two dozen value creation components based first on changes in equity value and then on changes in company capital structure, P&L, and market proxies. The order is important. Value creation formulas have the appearance of fundamental cause, but they are only backward-looking explanations of the real world measurement, which is equity value change.

The models in *Private Equity Value Creation Analysis* represent a few of several possible interpretations. For example, different formulas can produce different combinations of EBITDA Growth or Multiple Expansion that combine to provide identical $\Delta TEqV$ or $\Delta FEqV$ values. The interpretations in this volume are designed to be mathematically rigorous, rely on familiar and accessible input variables, and provide sensible and non-volatile results over the widest range of company, industry, and market conditions.

3.3. Accurate Equity Values

Since the analysis is, fundamentally, about equity value change, it is essential that all equity value input variables be accurate. Errors will flow through every value creation formula and shift value into the wrong components. Thus, shortcuts should never be taken with either *TEqV* or *FEqV* values.

This is usually not a problem for *FEqV* values, which are available in Fund reporting and GP marketing materials. However, *TEqV* values might not be reported, so analysts often resort to estimates based on other values. *TEqV* may be estimated from a known purchase price and funded debt at entry. Or it might be determined from a known *FEqV* value and an estimate of the Fund's ownership percentage. Both approaches can generate errors that materially change outcomes and interpretations.

For example, purchase price-driven *TEqV* values can be wrong if they exclude the impact of transaction fees, expenses, and other adjustments that occur before or after the transaction date. These may include working capital adjustments, retained cash, acquired liabilities, seller notes, etc. Ownership-driven $TEqV$ values can be wrong because, in each investment, Funds often own multiple securities with complex relationships that fluctuate with time and circumstance. Ownership values are often reported inconsistently, and sometimes incorrectly, making them the least reliable numbers in Fund reporting and GP marketing materials.

If $TEqV$ values are not available, analysts may need to examine equity waterfalls in company valuation files or the funds flow files for entry and exit. *Volume II: Applications* provides additional examples of how to gather and correctly format value creation input variables for numerous scenarios that are likely to be encountered in the real world. In practice, having a clear definition of $TEqV$ and explicitly asking the GP or deal team for numbers in the desired format is usually sufficient.

3.4. Directly Calculated Value Drivers

Some value creation models calculate one or more value creation components "indirectly". For example, the drivers of Fund-level value creation ($\Delta FEqV$) can be described as EBITDA Growth (V_{EG}), Multiple Expansion (V_{ME}), and the Leverage Effect (V_{LEV}). Knowing that $\Delta FEqV$ may be taken from Fund reporting materials, analysts might calculate V_{EG} and V_{ME} directly and then solve for V_{LEV} indirectly, as follows:

$$V_{LEV} = \Delta FEqV - V_{EG} - V_{ME}$$

This is a bad approach because any distortions in V_{EG} or V_{ME} will accumulate in V_{LEV}. Errors may be caused by incorrect $TEqV$ values (as illustrated above) or any number of improper adjustments to P&L or balance sheet numbers. Errors can also be driven by systemic biases in the formulas that favor transactions of certain types or sizes. While such biases may have nothing to do with leverage, they would always manifest themselves in V_{LEV}.

The following chapters show that every value driver can be calculated directly. The formula for V_{LEV} exists, and its values should be explicitly determined along with V_{EG} and V_{ME}. If these components do not sum to $\Delta FEqV$, then there is a problem with the formulas, input variables, or spreadsheet that must be addressed. Calculating every value driver directly helps to identify and correct such errors, making analyses more robust and reliable.

3.5. Accessible, Practical, and Non-volatile Input Variables

Value creation analyses require reasonable, representative, and reliable value creation input variables. If the analyst cannot obtain them, the most elegant or creative value creation formulas are of no use.

For example, an analytically interesting performance metric would be company free cash flow, but it is usually not provided in Fund reporting and GP marketing materials. Information about interest payment totals, effective tax rates and weighted average debt costs over a hold are not normally reported either, but they are required to calculate the impact of leverage in some value creation models (although not the ones recommended by this volume). Such metrics usually involve special requests to GPs or deal teams which limits how many and how frequently companies can be analyzed.

In contrast, Fund reporting and GP marketing materials often provide abbreviated company P&Ls with revenue, gross profit, and EBITDA values. These are more accessible input variables, so building the value creation formulas around them expands the universe of potential analysis candidates.

Practicality is another requirement. Input variables should be easy to update and not require significant "massaging" every quarter to make them meaningful. They should also be uncontroversial, so time and energy are not exhausted in defending their applicability. For example, value creation can be driven by growth in addressable market, but absolute ($) addressable market values are not practical input variables because they can be calculated a dozen different ways with results ranging from $50 million to $50 billion.

Finally, input variables must be non-volatile. Any big jump from one quarter to the next generates noise and spurious results. Company P&L metrics usually work well because TTM values change gradually. Any day's impact is small compared to a full year of revenue, gross profit, or EBITDA. Enterprise valuation metrics are another story. Public company valuation comparables can change by the minute as share prices change. Volatile input variables may require smoothing, perhaps in the form of a simple moving average, before being used in value creation models.

3.6. Mathematical Rigor

The last requirement is that value creation formulas be mathematically rigorous. This means that every value driver can be calculated directly, and values can be converted back and forth between the various measures like absolute ($) value creation ($V$), relative (%) value creation (v), equity return multiples or multipliers (χ), and value creation IRRs.

For example, Chapter 4 describes an investment's Unlevered Return as a combination of EBITDA Growth and Multiple Expansion. In a mathematically rigorous framework, measured values of V_{EG} and V_{ME} can be converted to v_{EG} and v_{ME}, and then to χ_{EG} and χ_{ME}, and then to IRR_{EG} and IRR_{ME}. These value creation IRRs combine to IRR_{UNLEV}, as follows:

$$IRR_{UNLEV} = (1 + IRR_{EG}) \cdot (1 + IRR_{ME}) - 1$$

Then IRR_{UNLEV} can be converted to χ_{UNLEV}, and then to v_{UNLEV}, and then to V_{UNLEV}, which should be identical to the initial sum of V_{EG} and V_{ME}.

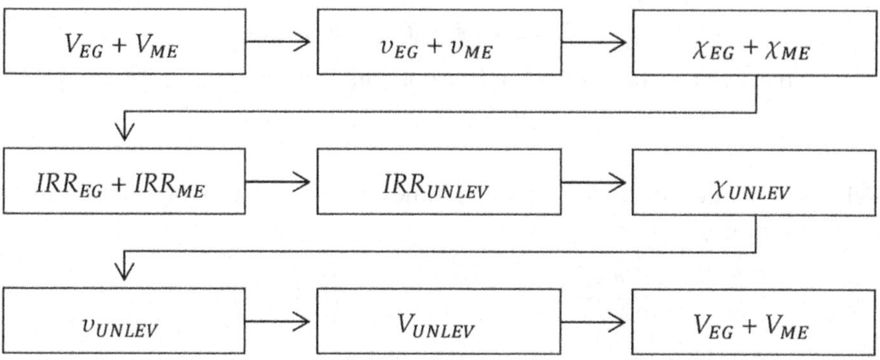

Figure 3.1 After eight transformations, $V_{EG} + V_{ME}$ should remain unchanged

The formulas derived in *Private Equity Value Creation Analysis* will insist on mathematical rigor. Some of the fundamental relationships underlying these derivations are as follows:

Additive Relationships

When a *Value* is the sum of components *a*, *b*, and *c*:

$$Value = a + b + c$$

Absolute ($) changes in *Value* between times t_1 and t_2 can be expressed as differences:

$$\begin{aligned} \Delta Value &= Value_2 - Value_1 \\ &= (a_2 + b_2 + c_2) - (a_1 + b_1 + c_1) \\ &= (a_2 - a_1) + (b_2 - b_1) + (c_2 - c_1) \\ &= \Delta a + \Delta b + \Delta c \\ &= V_a + V_b + V_c \end{aligned}$$

And relative (%) changes in *Value* between times t_1 and t_2 can be determined by dividing both sides by $\Delta\, Value$:

$$\frac{\Delta\, Value}{\Delta\, Value} = \frac{V_a + V_b + V_c}{\Delta\, Value} = 100\%$$

$$\frac{V_a}{\Delta\, Value} + \frac{V_b}{\Delta\, Value} + \frac{V_c}{\Delta\, Value} = 100\%$$

$$\frac{\Delta a}{\Delta\, Value} + \frac{\Delta b}{\Delta\, Value} + \frac{\Delta c}{\Delta\, Value} = 100\%$$

$$v_a + v_b + v_c = 100\%$$

Product Relationships

When a *Value* is the product of components x, y, and z:

$$Value = x \cdot y \cdot z$$

Relative changes in *Value* between times t_1 and t_2 can be expressed as multiples or multipliers (χ):

$$\frac{Value_2}{Value_1} = \frac{x_2 \cdot y_2 \cdot z_2}{x_1 \cdot y_1 \cdot z_1} = \left(\frac{x_2}{x_1}\right) \cdot \left(\frac{y_2}{y_1}\right) \cdot \left(\frac{z_2}{z_1}\right)$$

$$\chi_{Value} = \chi_x \cdot \chi_y \cdot \chi_z$$

When evaluating multipliers, natural logarithms can be particularly useful because they have the property that:

$$\ln(i \cdot j) = \ln(i) + \ln(j)$$

Therefore, relative (%) changes in *Value* between times t_1 and t_2 can be determined by taking the natural logarithm of χ_{Value} and then dividing by $\ln(\chi_{Value})$ to show that:

$$\frac{\ln(\chi_{Value})}{\ln(\chi_{Value})} = \frac{\ln\left(\chi_x \cdot \chi_y \cdot \chi_z\right)}{\ln(\chi_{Value})} = 100\%$$

$$\frac{\ln(\chi_x)}{\ln(\chi_{Value})} + \frac{\ln\left(\chi_y\right)}{\ln(\chi_{Value})} + \frac{\ln(\chi_z)}{\ln(\chi_{Value})} = 100\%$$

$$v_x + v_y + v_z = 100\%$$

Absolute ($) changes in *Value* between times t_1 and t_2 can be determined by multiplying both sides by $\Delta\, Value$ to provide:

$$\Delta\, Value = \Delta\, Value \cdot v_x + \Delta\, Value \cdot v_y + \Delta\, Value \cdot v_z$$

$$= \Delta\, Value \frac{\ln(\chi_x)}{\ln(\chi_{Value})} + \Delta\, Value \frac{\ln(\chi_y)}{\ln(\chi_{Value})} + \Delta\, Value \frac{\ln(\chi_z)}{\ln(\chi_{Value})}$$

$$= V_x + V_y + V_z$$

Because v_x, v_y, and v_z sum to 100%, χ_{Value} raised to the power of $v_x + v_y + v_z$ must equal χ_{Value}. A rule of exponents states that X raised to the power of a sum must be equal to the product of X raised to each component, so:

$$\chi_{Value}^{(v_x+v_y+v_z)} = \chi_{Value}^{v_x} \cdot \chi_{Value}^{v_y} \cdot \chi_{Value}^{v_z} = \chi_x \cdot \chi_y \cdot \chi_z$$

Multiplier and Growth Rate Relationships

Relative changes in *Value* between times t_1 and t_2 can be expressed as Compound Annual Growth Rates (CAGRs), using the multiple and holding period (Δt) measured in years:

$$CAGR_{Value} = \chi_{Value}^{1/\Delta t} - 1$$

By extension, multiples can be determined from CAGRs:

$$\chi_{Value} = (1 + CAGR_{Value})^{\Delta t}$$

Because χ_{Value} is the product of χ_x, χ_y, and χ_z, the relationship between the CAGR for *Value* and the CAGRs for x, y, and z is as follows:

$$\chi_{Value} = \chi_x \cdot \chi_y \cdot \chi_z$$

$$(1 + CAGR_{Value})^{\Delta t} = (1 + CAGR_x)^{\Delta t} \cdot \left(1 + CAGR_y\right)^{\Delta t} \cdot (1 + CAGR_z)^{\Delta t}$$

$$\left[(1 + CAGR_{Value})^{\Delta t} = (1 + CAGR_x)^{\Delta t} \cdot \left(1 + CAGR_y\right)^{\Delta t} \cdot (1 + CAGR_z)^{\Delta t}\right]^{1/\Delta t}$$

$$(1 + CAGR_{Value}) = (1 + CAGR_x) \cdot \left(1 + CAGR_y\right) \cdot (1 + CAGR_z)$$

The equity return multiplier for component x can be determined from χ_{Value} and the relative (%) change in x (v_x):

$$\chi_x = \chi_{Value}^{v_x}$$

Therefore, the CAGR for component x can be determined from χ_{Value}, v_x, and Δt, measured in years:

$$CAGR_x = \chi_{Value}^{v_x/\Delta t} - 1$$

3.7. Summary

The derivations of return measurements and value drivers in *Private Equity Value Creation Analysis* are based on these principles and best practices. They help to develop value creation models that are mathematically rigorous, rely on accessible data with which GPs and LPs are familiar, and provide sensible and non-volatile results over the widest range of company, industry, and market conditions. The next chapter turns to the "big three" value drivers identified by the most popular value creation models: EBITDA Growth, Multiple Expansion, and Leverage.

4. Conventional Model of Value Creation

This chapter describes the Conventional Model of value creation, which expresses company-level equity returns as the sum of EBITDA Growth (V_{EG}), Multiple Expansion (V_{ME}), and Cash Flow Generation (V_{CF}), where:

$$\Delta TEqV = V_{EG} + V_{ME} + V_{CF}$$
$$= \Delta E \cdot \overline{M} + \Delta M \cdot \overline{E} - \Delta ND$$

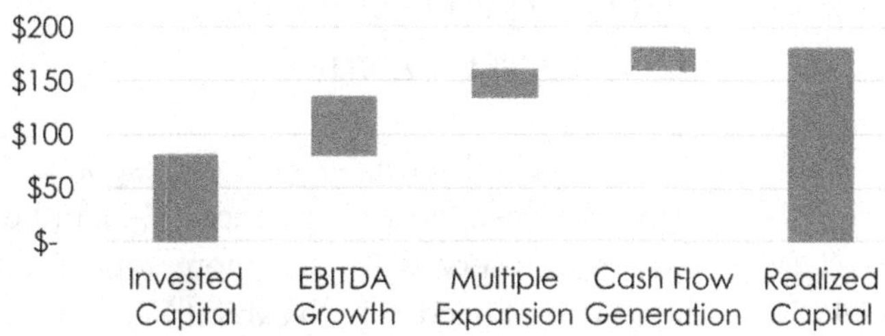

Figure 4.1 Basic value creation bridge

4.1. Background

The model is termed "Conventional" because LPs and GPs appear to use it more than any other. It is simple, robust, and intuitive. Returns and value drivers are calculated for all company shareholders and are generally expressed as absolute ($) value creation, with value drivers summing to equity returns. The analysis can be visualized as a value bridge or value gap analysis, like Figure 4.1, where value drivers "bridge the gap" between invested and realized capital.

The Conventional Model has been accused of being too simplistic and unsuitable for measuring company-specific and market-driven return components (White, Prahl, and Zeisberger 2016; Morris 2014), but this is overstated. Its formulas can be extended to produce every value driver described in *Private Equity Value Creation Analysis*. The model's only shortcoming is that it tends to understate the impact of leverage. For example, if the Conventional Model provides a Leverage/Market/Operational impact ratio of 30%:35%:35%, the other models might provide a ratio of 40%:30%:30%. The difference is meaningful, but in both cases the ratio of Market/Operational impact is 1:1. Depending on how the analysis is used, the leverage understatement may not be an issue. Furthermore, when Conventional Model value creation components are calculated as described in this chapter, they can be converted to the more sophisticated models with no loss of data integrity.

4.2. Conventional Model Derivation

Company-level total enterprise valuation (*TEV*) is the sum of its total equity value (*TEqV*) and net debt (*ND*), so *TEqV* can be expressed as $TEV - ND$. Company-level value creation between times t_1 and t_2 is the change in total equity value ($\Delta TEqV$), which provides:

$$\begin{aligned}\Delta TEqV &= TEqV_2 - TEqV_1 = (TEV_2 - ND_2) - (TEV_1 - ND_1) \\ &= (TEV_2 - TEV_1) - (ND_2 - ND_1) \\ &= \Delta TEV - \Delta ND\end{aligned}$$

The Conventional Model's second term, $-\Delta ND$, corresponds to balance sheet changes. This is defined as Cash Flow Generation (V_{CF}), rather than the Leverage Effect (V_{LEV}) which is reserved for the more sophisticated value creation models. Investments do not need debt to exhibit V_{CF} or V_{LEV}. Venture and growth investments may include excess cash to fund growth initiatives and, therefore, have negative *ND* values. In such instances, using the term net cash (*NC*) for the input variable, where $NC = -ND$, may help to avoid confusion.[1] Cash Flow Generation is therefore:

$$\begin{aligned}V_{CF} &= -(ND_2 - ND_1) = +(NC_2 - NC_1) \\ &= -\Delta ND = +\Delta NC\end{aligned}$$

The Conventional Model's first term, ΔTEV, is the change in company enterprise valuation. It is defined as the Unlevered Return (V_{UNLEV}) because in an all equity transaction *ND* would be \$0 and $\Delta TEqV$ would equal ΔTEV.

$$V_{UNLEV} = TEV_2 - TEV_1 = \Delta TEV$$

[1] Negative net debt values are sometimes described as \$0 or "n/a" in Fund reporting materials.

Central to value creation analyses are the assumptions that:

- Companies have enterprise valuations that are meaningful and variable.
- Valuations can be expressed as multiples of certain performance metrics.
- It is reasonable and informative to compare valuations, performance metrics, and valuation multiples at different times.

If the performance metric is EBITDA (E), the valuation multiple (M) would be TEV/E. With TEV expressed as $E \cdot M$, the Unlevered Return would be:

$$\begin{aligned}\Delta TEV &= TEV_2 - TEV_1 \\ &= E_2 \cdot M_2 - E_1 \cdot M_1\end{aligned}$$

This expression is accurate but useless because it does not express equity value change in terms of "value drivers". Useful value creation models must include terms like ΔE and ΔM that have the property of increasing company equity value whenever EBITDAs grow or valuation multiples expand.

A solution comes from basic calculus. Enterprise valuations, performance metrics, and valuation multiples are all functions of time, so at any time t:

$$TEV_{(t)} = E_{(t)} \cdot M_{(t)}$$

In physics, a change in position from point X_1 to X_2 is "displacement" (ΔX), and how fast displacement occurs can be expressed as ΔX divided by Δt, or "average velocity". Similarly, how fast enterprise valuations change can be expressed as ΔTEV divided by Δt. If the time interval Δt is allowed to become infinitesimally small, the rate of change approaches a limiting value known as the instantaneous rate of change, which can be expressed as a time derivative:

$$\lim_{\Delta t \to 0} \frac{\Delta TEV}{\Delta t} = \frac{\partial TEV_{(t)}}{\partial t}$$

Since performance metrics and valuation multiples are also functions of time, their relationship to the time derivative is described by the "Product Rule":

$$\frac{\partial TEV_{(t)}}{\partial t} = \frac{\partial}{\partial t}\left(E_{(t)} \cdot M_{(t)}\right) = \frac{\partial E_{(t)}}{\partial t} \cdot M_{(t)} + \frac{\partial M_{(t)}}{\partial t} \cdot E_{(t)}$$

This formula assumes that $TEV_{(t)}$, $E_{(t)}$, and $M_{(t)}$ are continuous and well-behaved functions. They probably are, but in practice their values are known only at certain times, like when a Fund quarterly report is published or when the company finance department closes its books. Replacing instantaneous changes (∂x) with discrete changes (Δx) provides:

$$\frac{\partial TEV_{(t)}}{\partial t} \approx \frac{\Delta TEV}{\Delta t} = \frac{\Delta E}{\Delta t} \cdot M_{(t)} + \frac{\Delta M}{\Delta t} \cdot E_{(t)}$$

And multiplying both sides by Δt provides:

$$\begin{aligned} \Delta TEV &= \Delta E \cdot M_{(t)} + \Delta M \cdot E_{(t)} \\ &= V_{EG} + V_{ME} \end{aligned}$$

This is the foundation of the Conventional Model. The $\Delta E \cdot M_{(t)}$ term represents the return driven by EBITDA Growth (V_{EG}) and the $\Delta M \cdot E_{(t)}$ term represents the return driven by Multiple Expansion (V_{ME}). Because $E_{(t)}$ and $M_{(t)}$ are generally positive numbers,[2] positive ΔE or ΔM values increase enterprise valuation, and negative ΔE or ΔM values decrease enterprise valuation.

So, what are $E_{(t)}$ and $M_{(t)}$? There are several mathematically valid options, including the following:

$$\begin{aligned} \Delta TEV_A &= \Delta E \cdot M_1 + \Delta M \cdot E_2 \\ \Delta TEV_B &= \Delta E \cdot M_2 + \Delta M \cdot E_1 \\ \Delta TEV_C &= \Delta E \cdot \frac{M_1 + M_2}{2} + \Delta M \cdot \frac{E_1 + E_2}{2} \\ \Delta TEV_D &= \Delta E \cdot M_1 + \Delta M \cdot E_1 + \Delta E \cdot \Delta M \\ \Delta TEV_E &= \Delta E \cdot M_2 + \Delta M \cdot E_2 - \Delta E \cdot \Delta M \end{aligned}$$

They are all mathematically valid because substituting ($E_2 - E_1$) for ΔE and ($M_2 - M_1$) for ΔM reduces every equation to $\Delta TEV = E_2 \cdot M_2 - E_1 \cdot M_1$, the original identity. Each formula produces different value drivers which sum to identical ΔTEV totals. None of the formulas are "wrong" but only one can be chosen, and that choice impacts every other value creation component because Market Share Growth is a function of Revenue Growth, which is a function of EBITDA Growth, which is impacted by EBITDA Margin Expansion, Multiple Expansion, Leverage, Fund Ownership Impact, etc.

Private Equity Value Creation Analysis recommends the use of ΔTEV_C, which uses the average of t_1 and t_2 EBITDA values for $E_{(t)}$ and the average of t_1 and t_2 valuation multiples for $M_{(t)}$. Average values are represented with an overbar:

---

[2] Negative EBITDA values are common but in such cases they are not used as a basis for valuation.

$$\overline{X} = \frac{X_1 + X_2}{2}$$

In fact, whenever $X_{(t)}$ appears in a value creation formula, this volume recommends that it be replaced with $\overline{X}$, as defined above. Other value creation models take different approaches, but the $\overline{X}$-formulations generally are more representative of overall investment conditions, they reduce volatility in results, and they provide additional flexibility and compatibility with value creation models described in later chapters. See Section 4.5 for additional detail on the use of averages as input variables in value creation formulas.

Conventional Absolute ($) Value Creation

Substituting ΔTEV_C into the company-level equity return formula ($\Delta TEqV$) describes absolute ($) value creation:

$$\begin{aligned} \Delta TEqV &= \Delta TEV_C - \Delta ND \\ &= \Delta E \cdot (M_1 + M_2)/2 + \Delta M \cdot (E_1 + E_2)/2 - \Delta ND \\ &= \Delta E \cdot \overline{M} + \Delta M \cdot \overline{E} - \Delta ND \\ &= V_{EG} + V_{ME} + V_{CF} \end{aligned}$$

Conventional Relative (%) Value Creation

Dividing both sides of the absolute ($) value creation formula by $\Delta TEqV$ describes relative (%) value creation:

$$\frac{\Delta TEqV}{\Delta TEqV} = \frac{V_{EG}}{\Delta TEqV} + \frac{V_{ME}}{\Delta TEqV} + \frac{V_{CF}}{\Delta TEqV} = 100\%$$

$$\frac{\Delta E \cdot \overline{M}}{\Delta TEqV} + \frac{\Delta M \cdot \overline{E}}{\Delta TEqV} + \frac{-\Delta ND}{\Delta TEqV} = 100\%$$

$$v_{EG} + v_{ME} + v_{CF} = 100\%$$

Conventional Equity Return Multipliers

The company equity return multiple (χ_{EQUITY}) is $TEqV_2/TEqV_1$. Raising χ_{EQUITY} to the power of each value driver's relative (%) value creation describes each value creation component's equity return multiplier:

$$\frac{TEqV_2}{TEqV_1} = \left(\frac{TEqV_2}{TEqV_1}\right)^{v_{EG}} \cdot \left(\frac{TEqV_2}{TEqV_1}\right)^{v_{ME}} \cdot \left(\frac{TEqV_2}{TEqV_1}\right)^{v_{CF}}$$

$$\chi_{EQUITY} = \chi_{EQUITY}{}^{v_{EG}} \cdot \chi_{EQUITY}{}^{v_{ME}} \cdot \chi_{EQUITY}{}^{v_{CF}}$$

$$= \chi_{EG} \cdot \chi_{ME} \cdot \chi_{CF}$$

Conventional Value Creation IRRs

The company equity IRR (IRR_{EQUITY}) can be estimated from χ_{EQUITY} and the investment hold (Δt) in years:

$$IRR_{EQUITY} = \chi_{EQUITY}{}^{1/\Delta t} - 1$$

Raising the equity return multiplier formula to the reciprocal of Δt describes the relationship between IRR_{EQUITY} and each component's value creation IRR:

$$\left(\chi_{EQUITY} = \chi_{EG} \cdot \chi_{ME} \cdot \chi_{CF}\right)^{1/\Delta t}$$

$$\chi_{EQUITY}{}^{1/\Delta t} = \chi_{EG}{}^{1/\Delta t} \cdot \chi_{ME}{}^{1/\Delta t} \cdot \chi_{CF}{}^{1/\Delta t}$$

$$\left(1 + IRR_{EQUITY}\right) = (1 + IRR_{EG}) \cdot (1 + IRR_{ME}) \cdot (1 + IRR_{CF})$$

4.3. Conventional Model Examples

Consider a company, AlphaCo, that is acquired for 5.0x EBITDA of \$20 at t_1 and sold for 6.0x EBITDA of \$30 at t_2. The transaction is funded with *ND* of \$20 at entry, which is paid down entirely at exit, providing:

Table 4.1. AlphaCo value creation input variables

	t_1	t_2	ΔX	$\overline{X}$
TEV	\$100	\$180	+\$80	-
TEqV	\$80	\$180	+\$100	-
ND	\$20	\$0	-\$20	-
E	\$20	\$30	+\$10	\$25
M	5.0x	6.0x	+1.0x	5.5x

Absolute (\$) value creation is calculated:

$$\Delta TEqV = V_{EG} + V_{ME} + V_{CF}$$

$$= \Delta E \cdot \overline{M} + \Delta M \cdot \overline{E} - \Delta ND$$

$$= \$10 \cdot 5.5x + 1.0x \cdot \$25 + \$20$$

$$= \$55 + \$25 + \$20 = \$100$$

Results may be visualized with a value gap or value bridge chart which shows that Invested Capital and the value creation components sum to Realized Capital.

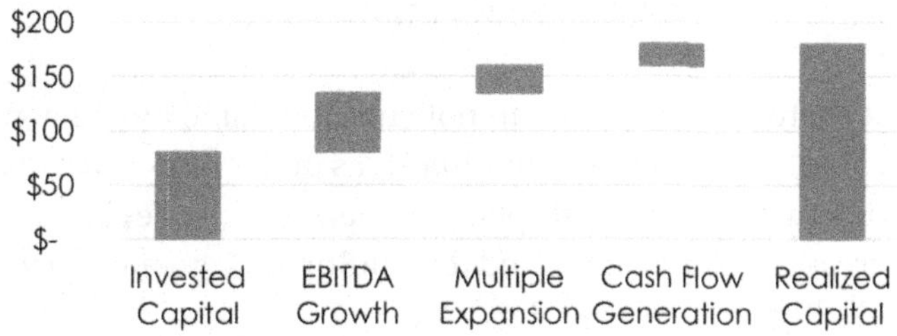

Figure 4.2 AlphaCo absolute ($) value creation

Relative (%) value creation is calculated:

$$v_{EG} + v_{ME} + v_{CF} = 100\%$$

$$\frac{V_{EG}}{\Delta TEqV} + \frac{V_{ME}}{\Delta TEqV} + \frac{V_{CF}}{\Delta TEqV} = 100\%$$

$$\frac{\$55}{\$100} + \frac{\$25}{\$100} + \frac{\$20}{\$100} = 100\%$$

$$55\% + 25\% + 20\% = 100\%$$

Results may be visualized with a stacked horizontal bar chart which illustrates how value drivers sum to return components. Below, EBITDA Growth and Multiple Expansion sum to Unlevered Return, and Cash Flow Generation and Unlevered Return sum to the Equity Return.

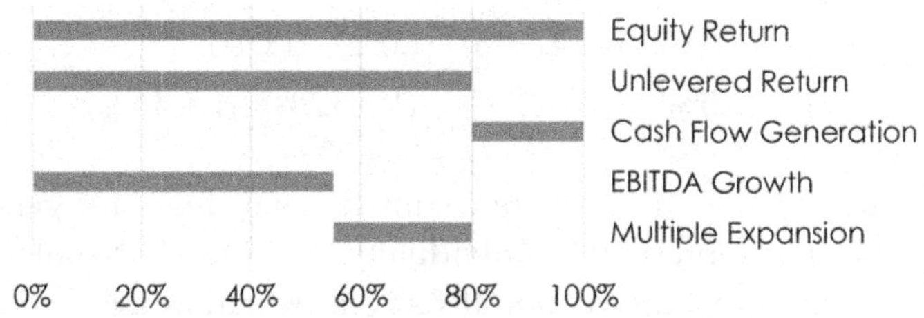

Figure 4.3 AlphaCo relative (%) value creation

The company equity multiple of invested capital (χ_{EQUITY}), determined by $TEqV_2/TEqV_1$, is 2.25x. The equity return multiplier for each value creation component is calculated:

$$\begin{aligned} x_{EQUITY} &= x_{EG} \cdot x_{ME} \cdot x_{CF} \\ &= x_{EQUITY}{}^{v_{EG}} \cdot x_{EQUITY}{}^{v_{ME}} \cdot x_{EQUITY}{}^{v_{CF}} \\ &= 2.25x^{0.55} \cdot 2.25x^{0.25} \cdot 2.25x^{0.20} \\ &= 1.56x \cdot 1.22x \cdot 1.18x = 2.25x \end{aligned}$$

Because equity return multipliers do not sum to total, they are generally not "stacked" and best represented individually, as in the vertical bar chart below. The impact of equity return multipliers is measured as deviation from 1.0x, with value creation being enhanced by multipliers greater than 1.0x and impaired by multipliers less than 1.0x.

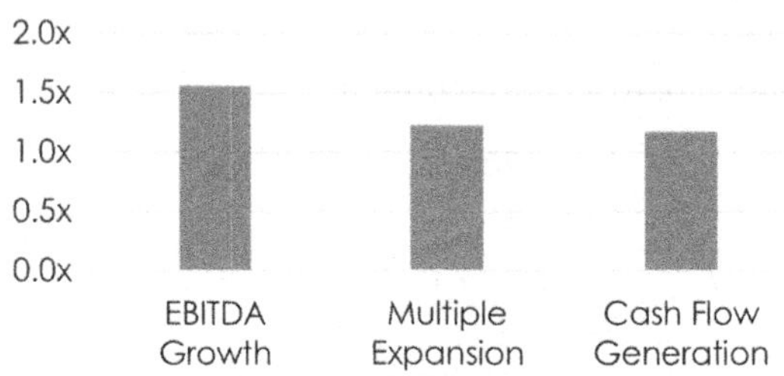

Figure 4.4 AlphaCo equity return multipliers

A company equity return multiple of 2.25x that is achieved in 4.0 years yields an equity IRR of 22.5%. The value creation IRR formula provides:

$$\left(1 + IRR_{EQUITY}\right) = (1 + IRR_{EG}) \cdot (1 + IRR_{ME}) \cdot (1 + IRR_{CF})$$

$$x_{EQUITY}{}^{1/\Delta t} = x_{EG}{}^{1/\Delta t} \cdot x_{ME}{}^{1/\Delta t} \cdot x_{CF}{}^{1/\Delta t}$$

$$2.25x^{1/4.0} = 1.56x^{1/4.0} \cdot 1.22x^{1/4.0} \cdot 1.18x^{1/4.0}$$

$$(1.225) = (1.118) \cdot (1.052) \cdot (1.041)$$

$$(1 + 22.5\%) = (1 + 11.8\%) \cdot (1 + 5.2\%) \cdot (1 + 4.1\%)$$

Because value creation IRRs do not sum to total, they are generally not "stacked" and best represented individually, as in the horizontal bar chart below. The impact of value creation IRRs is measured as deviation from 0%, with value creation being enhanced by positive value creation IRRs and impaired by negative value creation IRRs.

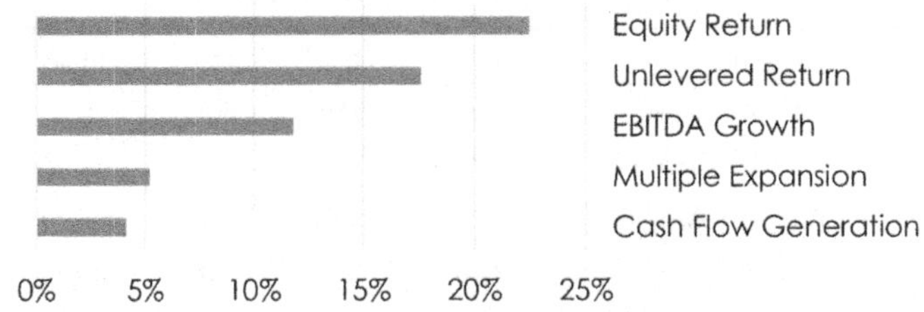

Figure 4.5 AlphaCo value creation IRRs

4.4. Summary

The Conventional Model is the most widely used form of value creation analysis. It states that company-level returns can be described as a combination of EBITDA Growth, Multiple Expansion, and Cash Flow Generation. The model can express returns and value drivers in the form of absolute ($) value creation, relative (%) value creation, equity return multiples or multipliers, and value creation IRRs.

The Conventional Model generally understates the impact of leverage. Cash Flow Generation is $-\Delta ND$, so if net debt does not change, then the Conventional Model states that debt has no impact on equity returns regardless of whether it makes up 0% or 80% of company capital structure. This seems unlikely since debt is known to enhance equity returns, at least when investments go well. The value creation models described in Chapter 5 examine leverage in greater detail and correct for this deficiency.

4.5. Optional: The Case for Holding Period Averages

Section 4.3 asserted that whenever $X_{(t)}$ appears in any value creation formula, it should be replaced with the average of X_1 and X_2 values ($\overline{X}$) because the $\overline{X}$-formulation is generally more representative of overall holding period conditions, it reduces volatility in value creation results, and it provides additional flexibility and compatibility with the value creation models described in later chapters. This "optional" section explores these claims.

First, consider the five Unlevered Return formulas that satisfy the condition $\Delta TEV = E_2 \cdot M_2 - E_1 \cdot M_1$. Each has terms for EBITDA Growth and Multiple Expansion. Two include an additional $\Delta E \cdot \Delta M$ term. Solving the components for AlphaCo provides:

Table 4.2. AlphaCo value creation alternatives

	Formula	V_{EG}	V_{ME}	Other	Total
$\Delta TEV_A =$	$\Delta E \cdot M_1 + \Delta M \cdot E_2$	\$50	\$30	-	\$80
$\Delta TEV_B =$	$\Delta E \cdot M_2 + \Delta M \cdot E_1$	\$60	\$20	-	\$80
$\Delta TEV_C =$	$\Delta E \cdot \overline{M} + \Delta M \cdot \overline{E}$	\$55	\$25	-	\$80
$\Delta TEV_D =$	$\Delta E \cdot M_1 + \Delta M \cdot E_1 + \Delta E \cdot \Delta M$	\$50	\$20	+\$10	\$80
$\Delta TEV_E =$	$\Delta E \cdot M_2 + \Delta M \cdot E_2 - \Delta E \cdot \Delta M$	\$60	\$30	–\$10	\$80

The ΔTEV_A and ΔTEV_B formulas have biases. If ΔE and ΔM are both positive or both negative, ΔTEV_A will favor Multiple Expansion and ΔTEV_B will favor EBITDA Growth. If ΔE and ΔM move in opposite directions, ΔTEV_A will favor EBITDA Growth and ΔTEV_B will favor Multiple Expansion. The ΔTEV_C formula will always "split the difference" between ΔTEV_A and ΔTEV_B, which reduces volatility in the value creation results.

The third terms in ΔTEV_D and ΔTEV_E are cross-products that represent the impact of changing EBITDA on changing valuation multiples. It is interesting because valuation multiples depend on EBITDA and, holding all other variables constant, higher EBITDA should demand a higher valuation. The cross-product is included in value creation models described by Capital Dynamics in several studies starting in 2009, where it is defined as a "Combo Factor" (Achleitner, Lichtner, and Diller 2009).

However, by incorporating one of these cross-products into the model, the analyst commits to creating a cross-product for every other value driver combination. There would also be cross-products for: Fund Ownership Impact and Isomeric Return, Unlevered Return and Leverage Effect, Revenue Growth and EBITDA Margin Expansion, Gross Margin Expansion and Operating Margin Reduction, Addressable Market Growth and Market Share Growth, Market Multiple Expansion and Intrinsic Multiple Expansion, etc. This would add at least seven cross-product terms to the value creation formulas and little to no analytical insight. Many of the cross-products are difficult to explain, or even comprehend.

A great benefit of the $\overline{X}$-formulations, like ΔTEV_C, is that they eliminate the noise of cross-products and mitigate the biases of variants like ΔTEV_A and ΔTEV_B. They also make more intuitive sense. AlphaCo's valuation multiple increases from 5.0x to 6.0x, so V_{ME} must be the product of 1.0x and some measure of holding period EBITDA ($\Delta M \cdot E_{(t)}$). Since AlphaCo's EBITDA grows from \$20 to \$30, one would expect the most representative $E_{(t)}$ value to fall somewhere between E_1 and E_2. The only EBITDA input in Table 4.2 that satisfies this criterion is the $\overline{E}$ of \$25 provided by ΔTEV_C.

To challenge this assertion, consider three real world scenarios with EBITDA growth profiles that exhibit: 1) slow growth, followed by acceleration; 2) fast growth, followed by a plateau; and 3) relatively steady growth.

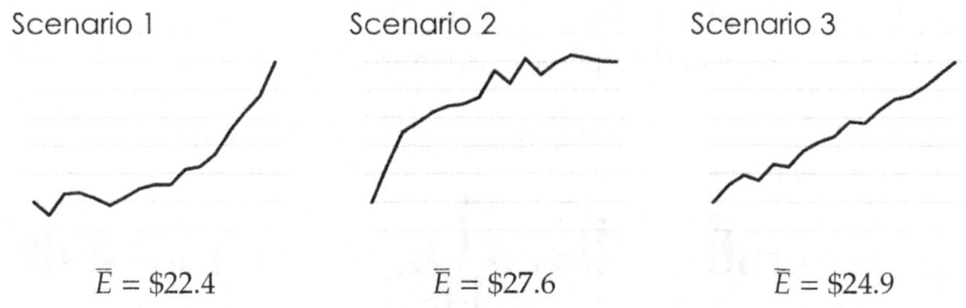

Figure 4.6 Holding period EBITDA growth scenarios for AlphaCo

For each scenario, $\overline{E}$ is the average of its 16 quarterly TTM EBITDAs. An analyst might be tempted to use these $\overline{E}$ values to solve for V_{ME}, assuming they provide a better "fit" to the holding period, but this would be a mistake. If additional precision is desired, a better approach is to break the 4.0-year hold into four one-year holds or 16 quarterly holds, and then calculate V_{ME} for each period. Holding period Multiple Expansion can then be modeled as the sum of all period values:

$$V_{ME} = \sum V_{ME(i)} = \sum \Delta M_i \cdot \overline{E_i}$$
$$= \Delta M_{1,2} \cdot \overline{E_{1,2}} + \Delta M_{2,3} \cdot \overline{E_{2,3}} + \Delta M_{3,4} \cdot \overline{E_{3,4}} + \cdots$$

This is like the calculus problem of Riemann Sums, where area under a curve is estimated by the sum of the areas of rectangles positioned at x_1, x_2, x_3, etc. Fit to the curve improves as Δx decreases and more rectangles are added. In this case, area under the curve is V_{ME} and fit to the curve improves as Δt decreases from one 4.0-year period, to four 1.0-year periods, and then to 16 three-month periods.

Fund reporting materials may provide enough information to complete such an analysis. Start with the Q1 and Q2 reports, turn to the one-page summaries for AlphaCo, note the difference in valuation multiples ($\Delta M_{1,2}$), take the average of TTM EBITDAs ($\overline{E_{1,2}}$), and then multiply them to determine $V_{ME(1,2)}$. Then toss Q1 to the side, pick up Q3, and repeat the process to determine $V_{ME(2,3)}$. Repeat 14 more times to provide 16 period $V_{ME(i)}$ values that add up to V_{ME}.

Figure 4.7 provides the $\Sigma V_{ME(i)}$ values for the three EBITDA growth profiles in Figure 4.6. Every scenario provides a $\Sigma V_{ME(i)}$ of \$25.4, despite exhibiting quite different EBITDA growth profiles. This suggests that the shape of EBITDA growth has no impact on Multiple Expansion. Furthermore, every scenario's result is quite close to the holding period $\Delta M \cdot \bar{E}$ value of \$25.0. This is noteworthy because it only took two quarterly reports to arrive at \$25.0, versus 17 quarterly reports to arrive at \$25.4.

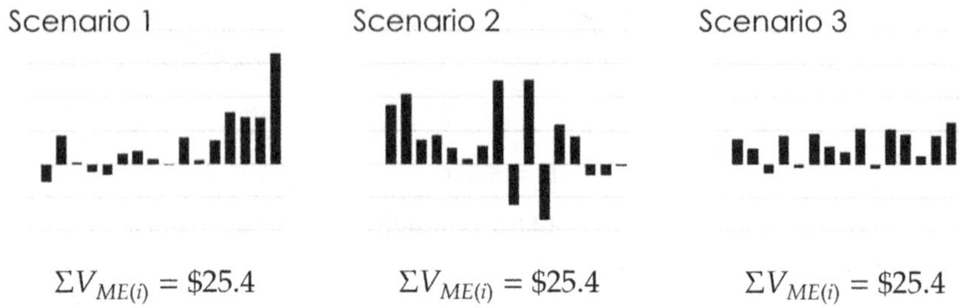

Figure 4.7 Quarterly Multiple Expansion values for each growth scenario

One might assume that the convergence toward \$25 is driven by the steps of $\Sigma V_{ME(i)}$ being modeled as $\Delta M \cdot \bar{E}$ (from ΔTEV_C), but this is not the case. Modeling each step as $\Delta M \cdot E_2$ (from ΔTEV_A) provides scenario $\Sigma V_{ME(i)}$ values of \$26.3, \$26.6, and \$25.9 (rather than \$30). Modeling each step as $\Delta M \cdot E_1$ (from ΔTEV_B) provides scenario $\Sigma V_{ME(i)}$ values of \$24.5, \$24.2, and \$24.9 (rather than \$20). Not only are the results generated by $\Delta M \cdot \bar{E}$ less volatile than those provided by $\Delta M \cdot E_1$ and $\Delta M \cdot E_2$, but as Δt decreases the $\Sigma V_{ME(i)}$ values for the later both converge to $\Delta M \cdot \bar{E}$. This supports the assertion that formulas like ΔTEV_C, which incorporate holding period averages, provide a better "fit" to real world conditions, in contrast to alternatives like ΔTEV_A, ΔTEV_B, ΔTEV_D, and ΔTEV_E.

The Multiple Expansion profiles in Figure 4.7 indicate that movements in $V_{ME(i)}$ are coordinated with changes in EBITDA. The $V_{ME(i)}$ gains are clustered at the end of Scenario 1's hold and at the beginning of Scenario 2's hold, where EBITDA changes are greatest. This makes sense if valuation is a function of EBITDA. Perhaps the $\Sigma V_{ME(i)}$ values are all \$25.4 because Multiple Expansion is not driven by the shape of EBITDA across time ($E_{(t)}$), but by the shape of valuation across EBITDA ($M_{(E)}$).

The $M_{(E)}$ formula can be estimated but never known precisely. For AlphaCo, it is known that $M_{(\$20)} = 5.0x$ because M_1 is 5.0x when E_1 is \$20, and that $M_{(\$30)} = 6.0x$ because M_2 is 6.0x when E_2 is \$30. There are infinite formulas that can satisfy these criteria, but one of them that is reasonable and "well-behaved" in the area around \$20 to \$30 EBITDA, is:

$$M_{(E)} = \frac{E^2}{400} - \frac{E}{40} + 4.5$$

This formula is used to determine $M_{(E)}$ values for the three scenarios above, which is why $\Sigma V_{ME(i)}$ for all of them converge to \$25.4. The expression can be used to calculate V_{EG} because $\Delta E \cdot M_{(E)}$ is effectively the area under the $M_{(E)}$ curve between an EBITDA of \$20 and \$30, such that:

$$V_{EG} = \int_{\$20}^{\$30} M_{(E)} dE$$

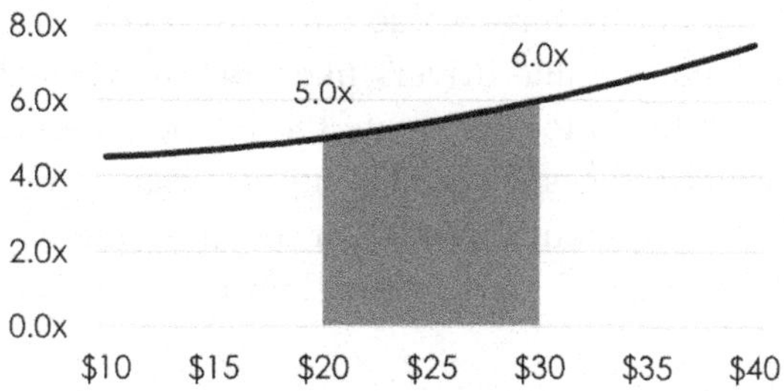

Figure 4.8 Integration of $M_{(E)}$

Area under the curve in Figure 4.8 can be determined precisely by solving the integral and taking its difference at EBITDA values of \$20 and \$30:

$$V_{EG} = \int_{\$20}^{\$30} M_{(E)} dE = \int_{\$20}^{\$30} \left(\frac{E^2}{400} - \frac{E}{40} + 4.5 \right) dE$$

$$= \left(\frac{E^3}{1200} - \frac{E^2}{80} + 4.5E + C \right) \Bigg|_{E=\$20}^{E=\$30}$$

$$= \frac{\$30^3}{1200} - \frac{\$30^2}{80} + 4.5 \cdot \$30 + C - \frac{\$20^3}{1200} + \frac{\$20^2}{80} - 4.5 \cdot \$20 - C = \$54.6$$

Integration of $M_{(E)}$ between EBITDAs of \$20 and \$30 provides V_{EG} of \$54.6. Additionally, recall that the Riemann Sum of $\Sigma V_{ME(i)}$ over any EBITDA growth profile that starts at \$20 and ends at \$30 provides V_{ME} of \$25.4. These values combine to provide the anticipated ΔTEV for AlphaCo:

$$\begin{aligned}\Delta TEV &= V_{EG} + V_{ME} \\ &= \$54.6 + 25.4 = \$80.0\end{aligned}$$

Analyses of this sort would never be performed for a real company. EBITDA growth profiles like those provided in Figure 4.6 generally are not available. Analysts certainly do not have access to the precise mathematical formulas describing the dependence of M on E. The point of the exercise is to demonstrate that value creation formulas which incorporate holding period averages get very close to whatever the "real" values should be. The analytical framework of *Private Equity Value Creation Analysis* relies on such formulas because they provide the most reasonable results, with the least information, over the widest range of scenarios.

Finally, only $\overline{X}$-formulations allow value creation results to be converted back and forth between the three value creation models. The Conventional Model usually overstates value drivers like EBITDA Growth and Multiple Expansion because it tends to understate the impact of leverage. However, when holding period averages are used as input variables, results can usually be converted to the Derivative Model or Logarithmic Model by simply multiplying the Conventional Model result by the company's average equity ratio ($\overline{ER}$).

5. Levered and Unlevered Returns

This chapter describes several ways to evaluate the impact of leverage on private company investment returns. Company-level equity returns are expressed as the sum of an Unlevered Return (V_{UNLEV}) and the Leverage Effect (V_{LEV}), where:

$$\begin{aligned} \Delta TEqV &= V_{UNLEV} + V_{LEV} \\ &= \Delta TEV \cdot \overline{ER} + \Delta ER \cdot \overline{TEV} \\ &= \Delta TEqV \frac{\ln(TEV_2/TEV_1)}{\ln(TEqV_2/TEqV_1)} + \Delta TEqV \frac{\ln(ER_2/ER_1)}{\ln(TEqV_2/TEqV_1)} \end{aligned}$$

5.1. Conceptual Model

The last chapter described the Unlevered Return as change in enterprise valuation. This means that when ΔND is \$0, the Conventional Model states that debt has no impact on equity returns, regardless of whether it makes up 0% or 80% of company capital structure.

Consider BravoCo, just like AlphaCo in Section 4.3, but with an interest-only loan that provides ND of \$50 at t_1 and t_2. Since BravoCo's enterprise valuation grows from \$100 to \$180, shareholder equity must grow from \$50 to \$130, generating $\Delta TEqV$ of \$80, an MOIC of 2.6x, and an IRR of 27.0% (assuming a 4.0-year hold).

Table 5.1 BravoCo value creation input variables

	t_1	t_2	ΔX
TEV	\$100	\$180	\$80
TEqV	\$50	\$130	\$80
ND	\$50	\$50	\$0

Consider the impact of leverage conceptually. Without the loan, shareholders must invest an extra \$50 of equity at t_1, only to have it returned at t_2. With a $TEqV_1$ of \$100 and a $TEqV_2$ of \$180, $\Delta TEqV$ would remain \$80, but because of the larger denominator the returns are reduced to an MOIC of 1.8x and IRR of 15.8%. Actually, the Unlevered Return is a bit better than this because the loan's interest payments would remain on BravoCo's balance sheet and, after taxes, increase company equity value. If the interest rate is 5%, then four annual interest payments sum to \$10, perhaps \$7 after tax, and provide $\Delta TEqV$ of \$87, an MOIC of 1.87x, and IRR of 16.9%.

This "Conceptual Model" can be generalized by defining the unlevered equity return multiple (χ_{UNLEV}) as the sum of TEV_2 and all after tax interest payments, divided by TEV_1:

$$\chi_{UNLEV} = \frac{TEV_2 + \sum(1 - T_i) \cdot Interest_i}{TEV_1} = \frac{TEV_2 + \sum(1 - T_i) \cdot r_D \cdot Debt_i}{TEV_1}$$

The Σ-term represents the sum of every interest payment after taxes (T_i). Interest payments can be represented as products of each period's weighted average cost of debt capital (r_D) and debt level ($Debt_i$). However, instead of calculating values for every period, the Σ-term can be estimated by multiplying t_1 and t_2 average values by the hold (Δt) in years:

$$\chi_{UNLEV} = \frac{TEV_2 + (1 - \overline{T}) \cdot \overline{r_D} \cdot \overline{ND} \cdot \Delta t}{TEV_1}$$

The Unlevered IRR (IRR_{UNLEV}) can be determined from χ_{UNLEV}:

$$IRR_{UNLEV} = \chi_{UNLEV}{}^{1/\Delta t} - 1 = \left(\frac{TEV_2 + (1 - \overline{T}) \cdot \overline{r_D} \cdot \overline{ND} \cdot \Delta t}{TEV_1}\right)^{1/\Delta t} - 1$$

Solving for BravoCo provides:

$$IRR_{UNLEV} = \left(\frac{\$180 + (1 - 30\%) \cdot 5.0\% \cdot \$50 \cdot 4.0}{\$100}\right)^{1/4.0} - 1 = 16.9\%$$

It is sometimes assumed that with a company equity IRR (IRR_{EQUITY}) of 27.0% and an IRR_{UNLEV} of 16.9%, the Leverage Effect (IRR_{LEV}) should be the difference, or 10.1%. In fact, it is 8.6% because value creation IRR components are related to each other through multiplication rather than addition:

$$\left(1 + IRR_{EQUITY}\right) = (1 + IRR_{UNLEV}) \cdot (1 + IRR_{LEV})$$

$$(1 + 27.0\%) = (1 + 16.9\%) \cdot (1 + 8.6\%)$$

5.2. WACC Model

In practice, IRR is something miraculously generated by a spreadsheet. It can be solved on paper if an investment has only two or three cash flows. With more cash flows it becomes an iterative calculation that requires a CPU. IRR is technically the discount rate that makes the present value of all investment cash flows equal to $0. Another popular discount rate in finance is the Weighted Average Cost of Capital, or WACC. It is generally used to measure an organization's financing costs for project decisions and valuation purposes, but WACC's form points to another solution for measuring the impact of leverage in private company investment returns:

$$WACC = \left(\frac{Equity}{Equity + Debt}\right) \cdot r_E + \left(\frac{Debt}{Equity + Debt}\right) \cdot (1 - T) \cdot r_D$$

The first term is the cost of equity (r_E) multiplied by the percentage of equity in the capital structure, or equity ratio (*ER*). The second term is the after-tax cost of debt (r_D) multiplied by the percentage of debt in the capital structure, or Debt Ratio (*DR*). Since each variable can change over time, t_1 and t_2 average values are generally used, providing:

$$WACC = \overline{ER} \cdot \overline{r_E} + \overline{DR} \cdot (1 - \overline{T}) \cdot \overline{r_D}$$

Compare the mechanics of WACC to the Conceptual Model's IRR_{UNLEV}. For an unlevered company, both $\overline{ND}$ and $\overline{DR}$ equal zero, so $IRR_{UNLEV} = IRR_{EQUITY}$ and $WACC = r_E$. For levered investments, $\overline{ND}$ and $\overline{DR}$ increase or decrease together and tend to shift IRR_{UNLEV} in the same direction as WACC, and IRR_{EQUITY} in the same direction as r_E. If IRR_{EQUTIY} is somehow equivalent to $\overline{r_E}$, and if IRR_{UNLEV} is somehow equivalent to WACC, then:

$$IRR_{UNLEV} = \overline{ER} \cdot IRR_{EQUITY} + \overline{DR} \cdot (1 - \overline{T}) \cdot \overline{r_D}$$

This is the WACC Model of Unlevered IRR. It is similar to a model described by Capital Dynamics in several studies starting in 2009 (Achleitner, Lichtner, and Diller 2009). In fact, when the WACC Model's tax discount is ignored ($\overline{T} = 0\%$), it can be shown that the WACC Model is identical to the following formulation used in the Capital Dynamics papers:

$$IRR_{EQUITY} = IRR_{UNLEV} + (IRR_{UNLEV} - \overline{r_D}) \cdot \overline{\left(\frac{Debt}{Equity}\right)}$$

Updating BravoCo's value creation input variables to include *ER* (*TEqV* divided by *TEV*) and *DR* (*ND* divided by *TEV*), and then substituting values into the WACC Model's Unlevered IRR formula provides:

Table 5.2 BravoCo value creation input variables

	t_1	t_2	ΔX	$\overline{X}$
TEV	\$100	\$180	+\$80	\$140
TEqV	\$50	\$130	+\$80	\$90
ND	\$50	\$50	\$0	\$50
ER	50%	72.2%	+22.2%	61.1%
DR	50%	27.8%	-22.2%	38.9%

$$IRR_{UNLEV} = \overline{ER} \cdot IRR_{EQUITY} + \overline{DR} \cdot (1 - \overline{T}) \cdot \overline{r_D}$$
$$= 61.1\% \cdot 27.0\% + 38.9\% \cdot (1 - 30\%) \cdot 5.0\%$$
$$= 16.5\% + 1.4\% = 17.9\%$$

Recall that the Conceptual Model first calculates a baseline Unlevered IRR of 15.8%, and then adjusts it upward by 1.1% to account for retained after-tax interest payments. The WACC Model appears to have similar mechanics.

The WACC Model would provide an IRR_{LEV} of 7.7% for BravoCo:

$$\left(1 + IRR_{EQUITY}\right) = (1 + IRR_{UNLEV}) \cdot (1 + IRR_{LEV})$$
$$(1 + 27.0\%) = (1 + 17.9\%) \cdot (1 + 7.7\%)$$

5.3. Accessibility of Input Variables

The Conceptual and WACC Models both rely on value creation input variables that generally are not included in Fund reporting and GP marketing materials. Interest payment totals, effective tax rates, and weighted average debt costs would involve special requests to GPs or deal teams. Most GPs would be happy to provide this detail for a few companies. Some would be willing to provide it on an ongoing basis for new investments. However, it would be a significant task to compile all the necessary data for a large portfolio, especially if it includes many historical companies that are no longer managed by the GP. These models, therefore, fall short of the principle outlined in Section 3.5 that input variables should be accessible.

Fortunately, the value creation models described in the next two sections achieve similar results for BravoCo, and require no details about interest payment totals, effective tax rates, or weighted average debt costs.

5.4. Derivative Model

The presence of the equity ratio in the WACC Model suggests another angle of attack. Analyses like the DuPont Formula use various ratios, including equity ratios, to break public company equity returns into various value drivers. Starting with company $TEqV$ and multiplying by 1.0x in the form of TEV/TEV provides:

$$TEqV = TEqV \cdot \left(\frac{TEV}{TEV}\right) = TEV \cdot \left(\frac{TEqV}{TEV}\right) = TEV \cdot ER$$

$TEqV$, TEV, and ER are all functions of time, so at any time t:

$$TEqV_{(t)} = TEV_{(t)} \cdot ER_{(t)}$$

These terms can be expressed as time derivatives that are related by the "Product Rule":[3]

$$\frac{\partial TEqV_{(t)}}{\partial t} = \frac{\partial}{\partial t}\left(TEV_{(t)} \cdot ER_{(t)}\right) = \frac{\partial TEV_{(t)}}{\partial t} \cdot ER_{(t)} + \frac{\partial ER_{(t)}}{\partial t} \cdot TEV_{(t)}$$

Replacing instantaneous changes (∂x) with discrete changes (Δx) and then multiplying both sides by Δt provides:

$$\begin{aligned} \Delta TEqV &= \Delta TEV \cdot \overline{ER} + \Delta ER \cdot \overline{TEV} \\ &= V_{UNLEV} + V_{LEV} \end{aligned}$$

This is the foundation of the Derivative Model. The $\Delta TEV \cdot \overline{ER}$ term is the Unlevered Return (V_{UNLEV}). It is the change in company enterprise valuation multiplied by the average percentage of equity in the company capital structure. For a levered company, $\overline{ER}$ is less than 100%. Higher debt drives lower $\overline{ER}$ values and lowers Unlevered Returns. The lack of this $\overline{ER}$-term in Chapter 4 explains why the Conventional Model tends to understate the impact of leverage. The $\Delta ER \cdot \overline{TEV}$ term is the Leverage Effect (V_{LEV}). Even though BravoCo's absolute ($) debt level does not change, it exhibits a positive Leverage Effect because its equity ratio increases 22.2% as enterprise valuation grows from $100 to $180.

Table 5.2 above provides the necessary value creation input variables for BravoCo. Solving absolute ($) value creation provides:

[3] A similar derivation, with additional steps and detail, is demonstrated in Section 4.2.

$$\begin{aligned}\Delta TEqV &= V_{UNLEV} + V_{LEV}\\ &= \Delta TEV \cdot \overline{ER} + \Delta ER \cdot \overline{TEV}\\ &= \$80 \cdot 61.1\% + 22.2\% \cdot \$140\\ &= \$48.9 + \$31.1 = \$80.0\end{aligned}$$

Relative (%) value creation can be determined from absolute ($) value creation:

$$\begin{aligned}&v_{UNLEV} + v_{LEV} = 100\%\\ &\frac{V_{UNLEV}}{\Delta TEqV} + \frac{V_{LEV}}{\Delta TEqV} = 100\%\\ &\frac{\$48.89}{\$80} + \frac{\$31.11}{\$80} = 100\%\\ &61.1\% + 38.9\% = 100\%\end{aligned}$$

Equity return multipliers can be determined from relative (%) value creation:

$$\begin{aligned}X_{EQUITY} &= X_{UNLEV} \cdot X_{LEV}\\ &= X_{EQUITY}{}^{v_{UNLEV}} \cdot X_{EQUITY}{}^{v_{LEV}}\\ &= 2.60x^{0.611} \cdot 2.60x^{0.389}\\ &= 1.79x \cdot 1.45x = 2.60x\end{aligned}$$

Value creation IRRs can be determined from equity return multipliers:

$$\begin{gathered}\left(1 + IRR_{EQUITY}\right) = (1 + IRR_{UNLEV}) \cdot (1 + IRR_{LEV})\\ X_{EQUITY}{}^{1/\Delta t} = X_{UNLEV}{}^{1/\Delta t} \cdot X_{LEV}{}^{1/\Delta t}\\ 2.6x^{1/4.0} = 1.79x^{1/4.0} \cdot 1.45x^{1/4.0}\\ (1.270) = (1.157) \cdot (1.097)\\ (1 + 27.0\%) = (1 + 15.7\%) \cdot (1 + 9.7\%)\end{gathered}$$

The Derivative Model provides an IRR_{UNLEV} of 15.7%, which is 1.2% lower than the Conceptual Model's 16.9%. The results are achieved without tax rate or interest payment details required by the Conceptual and WACC models.

5.5. Logarithmic Model

The Logarithmic Model recognizes that if company equity value is the product of enterprise valuation and equity ratio, then changes in company equity value can be expressed as equity return multipliers:

$$\frac{TEqV_2}{TEqV_1} = \frac{(TEV_2 \cdot ER_2)}{(TEV_1 \cdot ER_1)} = \frac{TEV_2}{TEV_1} \cdot \frac{ER_2}{ER_1}$$

$$\chi_{EQUITY} = \chi_{UNLEV} \cdot \chi_{LEV}$$

This is the foundation of the Logarithmic Model. The company equity return multiple (χ_{EQUITY}) equals the product of an Unlevered Return multiple (χ_{UNLEV}) and Leverage Effect multiplier (χ_{LEV}). BravoCo's return multipliers can be calculated from the input variables above:

Table 5.3 BravoCo value creation input variables

	t_1	t_2	X_2/X_1
TEV	\$100	\$180	1.80x
TEqV	\$50	\$130	2.60x
ND	\$50	\$50	n/a
ER	50%	72.2%	1.44x

$$\chi_{EQUITY} = \chi_{UNLEV} \cdot \chi_{LEV}$$
$$= 1.80x \cdot 1.44x = 2.60x$$

These values are almost identical to the Derivative Model's equity return multipliers. Value creation IRRs can be determined from equity return multipliers:

$$\left(1 + IRR_{EQUITY}\right) = (1 + IRR_{UNLEV}) \cdot (1 + IRR_{LEV})$$

$$\chi_{EQUITY}{}^{1/\Delta t} = \chi_{UNLEV}{}^{1/\Delta t} \cdot \chi_{LEV}{}^{1/\Delta t}$$

$$2.60x^{1/4.0} = 1.80x^{1/4.0} \cdot 1.44x^{1/4.0}$$

$$(1.270) = (1.158) \cdot (1.096)$$

$$(1 + 27.0\%) = (1 + 15.8\%) \cdot (1 + 9.6\%)$$

The Logarithmic Model provides an IRR_{UNLEV} of 15.8%, which is 1.1% lower than the Conceptual Model's 16.9% and, again, very close to the Derivative Model value. The following chapters will show that the Logarithmic Model tends to arrive at equity return multipliers and value creation IRRs with less data and fewer steps than any other model.

Logarithmic Model measurements of relative (%) value creation and absolute (\$) value creation require a few more steps. The continuous change in company equity value (r_{EQUITY}) between times t_1 and t_2 may be expressed:

$$r_{EQUITY} = \ln(TEqV_2) - \ln(TEqV_1) = \ln\left(\frac{TEqV_2}{TEqV_1}\right)$$

Since $TEqV_2/TEqV_1$ is χ_{EQUITY}, the continuous change is:

$$r_{\text{EQUITY}} = \ln(\chi_{EQUITY}) = \ln(\chi_{UNLEV} \cdot \chi_{LEV})$$

Natural logarithms have the useful property that $\ln(a \cdot b)$ equals the sum of $\ln(a)$ and $\ln(b)$, so the continuous change in company equity value can be expressed:

$$r_{\text{EQUITY}} = \ln(\chi_{EQUITY}) = \ln(\chi_{UNLEV}) + \ln(\chi_{LEV}) = r_{\text{UNLEV}} + r_{\text{LEV}}$$

The r_{EQUITY} expression suggests that the continuous change in company equity value is equal to the sum of the continuous changes in enterprise valuation (r_{UNLEV}) and equity ratio (r_{LEV}). Dividing both sides by $\ln(\chi_{EQUITY})$ provides:

$$\frac{\ln(\chi_{EQUITY})}{\ln(\chi_{EQUITY})} = \frac{\ln(\chi_{UNLEV})}{\ln(\chi_{EQUITY})} + \frac{\ln(\chi_{LEV})}{\ln(\chi_{EQUITY})} = 100\%$$

$$v_{UNLEV} + v_{LEV} = 100\%$$

This is a useful way to measure company relative (%) value creation. It is not as intuitive as the Conventional or Derivative models, but it is often the fastest way to produce a result, especially when many value creation components are involved.

Solving for BravoCo provides:

$$v_{UNLEV} + v_{LEV} = 100\%$$

$$\frac{\ln(1.80x)}{\ln(2.60x)} + \frac{\ln(1.44x)}{\ln(2.60x)} = 100\%$$

$$61.5\% + 38.5\% = 100\%$$

Multiplying both sides of the relative (%) value creation formula by $\Delta TEqV$ provides a solution for absolute ($) value creation:

$$\Delta TEqV = V_{UNLEV} + V_{LEV}$$

$$= v_{UNLEV} \cdot \Delta TEqV + v_{LEV} \cdot \Delta TEqV$$

$$= 61.5\% \cdot \$80 + 38.5\% \cdot \$80$$

$$= \$49.2 + \$30.8 = \$80.0$$

While none of the Logarithmic Model values are identical to Derivative Model values, they are all generally within a few tenths of a percent.

5.6. Model Comparisons

BravoCo's value creation IRRs, provided by each model, are given below:

Table 5.4 Levered and Unlevered Return results

	IRR_{UNLEV}	IRR_{LEV}	IRR_{EQUITY}
Conventional Model	27.0%	n/a	27.0%
Conceptual Model	16.9%	8.6%	27.0%
WACC Model	17.9%	7.7%	27.0%
Derivative Model	15.7%	9.7%	27.0%
Logarithmic Model	15.8%	9.6%	27.0%

If the Conceptual Model described in Section 5.1 is assumed to provide the "true" representation of Unlevered Return and Leverage Effect, the WACC Model seems to have the smallest "error" and perform the best. However, this may be valid for only BravoCo. The models should be tested across various capital structure and return scenarios.

In the following examples, TEV_1 is fixed at \$100, and TEV_2 is fixed at \$180. The ND_1 values vary from \$0 to \$75, and the ND_2 values vary from \$0 to \$175. Equity value variables are calculated assuming that $TEqV_1 = \$100 - ND_1$ and $TEqV_2 = \$180 - ND_2$. Changing net debt in increments of \$25 provides 32 different return scenarios for each of the four models, and a total of 128 IRR_{UNLEV} measurements.

The Conceptual Model provides:

$$IRR_{UNLEV} = \left(\frac{TEV_2 + (1 - \overline{T}) \cdot \overline{r_D} \cdot \overline{ND} \cdot \Delta t}{TEV_1} \right)^{1/\Delta t} - 1$$

$$= \left(\frac{\$180 + (1 - 30\%) \cdot 5.0\% \cdot \overline{ND} \cdot 4.0}{\$100} \right)^{1/4.0} - 1$$

Table 5.5 Conceptual Model Unlevered IRRs

		ND_2							
		\$0	\$25	\$50	\$75	\$100	\$125	\$150	\$175
ND_1	\$0	*15.8%*	*16.1%*	*16.4%*	*16.7%*	*16.9%*	*17.2%*	*17.5%*	*17.8%*
	\$25	*16.1%*	*16.4%*	*16.7%*	*16.9%*	*17.2%*	*17.5%*	*17.8%*	*18.0%*
	\$50	*16.4%*	*16.7%*	*16.9%*	*17.2%*	*17.5%*	*17.8%*	*18.0%*	*18.3%*
	\$75	*16.7%*	*16.9%*	*17.2%*	*17.5%*	*17.8%*	*18.0%*	*18.3%*	*18.5%*

As expected, the baseline IRR_{UNLEV} is 15.8% in the debt-free scenario, and higher debt levels modestly increase IRR_{UNLEV} through hypothetical retained after tax interest payments.

One might expect similar results from the WACC Model, which has comparable mechanics, but its IRR_{UNLEV} values are quite different:

$$IRR_{UNLEV} = \overline{ER} \cdot IRR_{EQUITY} + \overline{DR} \cdot (1 - \overline{T}) \cdot \overline{r_D}$$

$$= \overline{ER} \cdot \left(\left(\frac{\$180 - ND_2}{\$100 - ND_1} \right)^{1/4.0} - 1 \right) + \overline{DR} \cdot (1 - 30\%) \cdot 5.0\%$$

Table 5.6 WACC Model Unlevered IRRs

		ND_2							
		\$0	\$25	\$50	\$75	\$100	\$125	\$150	\$175
ND_1	\$0	*15.8%*	*11.0%*	*6.3%*	*1.7%*	*-2.9%*	*-7.8%*	*-13.7%*	*-25.4%*
	\$25	*21.8%*	*16.7%*	*11.8%*	*7.0%*	*2.4%*	*-2.3%*	*-7.5%*	*-17.0%*
	\$50	*29.2%*	*23.4%*	*17.9%*	*12.6%*	*7.7%*	*3.1%*	*-1.7%*	*-9.0%*
	\$75	*41.2%*	*33.7%*	*26.6%*	*20.0%*	*14.0%*	*8.6%*	*3.7%*	*-1.6%*

In only two scenarios, where $ND_1 = ND_2$ at \$25 and \$50, does the WACC Model outperform the Derivative and Logarithmic models. Of the 32 scenarios, only three have errors within 2%, and only seven have errors within 5%. The largest error is 43.1%, and the 32 scenarios have an average absolute error of 14.5% and a median absolute error of 13.7%.

The Derivative Model improves the analysis significantly:

$$IRR_{UNLEV} = \left(\frac{TEqV_2}{TEqV_1} \right)^{\frac{\Delta TEV \cdot \overline{ER}}{\Delta TEqV \cdot \Delta t}} - 1 = \left(\frac{\$180 - \mathrm{ND}_2}{\$100 - \mathrm{ND}_1} \right)^{\frac{\$80 \cdot \overline{ER}}{(\$80 - \Delta ND) \cdot 4.0}} - 1$$

Table 5.7 Derivative Model Unlevered IRRs

		ND_2							
		\$0	\$25	\$50	\$75	\$100	\$125	\$150	\$175
ND_1	\$0	*15.8%*	*16.0%*	*16.3%*	*16.7%*	*17.5%*	*18.9%*	*22.2%*	*38.3%*
	\$25	*15.7%*	*15.7%*	*15.9%*	*16.1%*	*16.7%*	*17.8%*	*20.5%*	*35.1%*
	\$50	*15.9%*	*15.8%*	*15.7%*	*15.7%*	*15.9%*	*16.6%*	*18.6%*	*31.0%*
	\$75	*17.3%*	*16.9%*	*16.5%*	*16.1%*	*15.8%*	*15.7%*	*16.4%*	*25.0%*

Of the 32 Derivative Model scenarios, 25 have errors within 2%, and 28 have errors within 5%. The largest error is 20.5%, and the 32 scenarios have an average absolute error of 2.7% and a median absolute error of 0.84%. The Derivative Model appears to be stressed only in scenarios with large ΔND values and debt ratios over 80% at exit.

The Logarithmic Model also outperforms the WACC Model, and does so with the simplest formula of all:

$$IRR_{UNLEV} = \left(\frac{TEV_2}{TEV_1}\right)^{1/\Delta t} - 1 = \left(\frac{\$180}{\$100}\right)^{1/4.0} - 1 = 15.8\%$$

Table 5.8 Logarithmic Model Unlevered IRRs

		ND_2							
		\$0	\$25	\$50	\$75	\$100	\$125	\$150	\$175
ND_1	\$0	*15.8%*	*15.8%*	*15.8%*	*15.8%*	*15.8%*	*15.8%*	*15.8%*	*15.8%*
	\$25	*15.8%*	*15.8%*	*15.8%*	*15.8%*	*15.8%*	*15.8%*	*15.8%*	*15.8%*
	\$50	*15.8%*	*15.8%*	*15.8%*	*15.8%*	*15.8%*	*15.8%*	*15.8%*	*15.8%*
	\$75	*15.8%*	*15.8%*	*15.8%*	*15.8%*	*15.8%*	*15.8%*	*15.8%*	*15.8%*

All IRR_{UNLEV} values are the same because the Logarithmic Model only relies on enterprise valuations, which are fixed for this exercise. It is effectively the Conceptual Model's baseline Unlevered Return, with no adjustment for hypothetical retained interest payments. In this case, the shortcut does not harm relative accuracy. Of the 32 Logarithmic Model scenarios, 28 have errors within 2%, and the largest error is 2.7%. The 32 scenarios have an average absolute error of 1.4% and a median absolute error of 1.4%.

Does this make the Logarithmic Model the best approach? It depends. Of the 32 scenarios, the Derivative Model is closest to the Conceptual Model 17 times, the Logarithmic Model is closest 12 times, and the WACC Model is closest twice. All four models are identical when $ND_1 = ND_2 = \$0$.

Table 5.9 Model producing results closest to the Conceptual Model

		ND_2							
		\$0	\$25	\$50	\$75	\$100	\$125	\$150	\$175
ND_1	\$0	-	*Deriv.*	*Deriv.*	*Deriv.*	*Deriv.*	*Log.*	*Log.*	*Log.*
	\$25	*Log.*	*WACC*	*Deriv.*	*Deriv.*	*Deriv.*	*Deriv.*	*Log.*	*Log.*
	\$50	*Deriv.*	*Log.*	*WACC*	*Log.*	*Deriv.*	*Deriv.*	*Deriv.*	*Log.*
	\$75	*Deriv.*	*Deriv.*	*Deriv.*	*Deriv.*	*Log.*	*Log.*	*Deriv.*	*Log.*

5.7. Summary

The Derivative and Logarithmic models of value creation provide reasonable measures of Unlevered Return and Leverage Effect. Both models can express returns and value drivers in the form of absolute (\$) value creation, relative (%) value creation, equity return multiples or multipliers, and value creation IRRs. Results are achieved without cash flow models or company specific input variables that can be difficult to obtain or verify, like interest payment totals, effective tax rates, or weighted average debt costs.

The value creation components provided by the Derivative and Logarithmic models are generally close enough that they are almost interchangeable. This is true for most value drivers, not just those associated with leverage. That said, it is generally best to stick to one of the two models to maintain consistency across analyses.

The next chapter revisits the Conventional Model's description of EBITDA Growth and Multiple Expansion in the context of these improved leverage models.

5.8. Optional: Derivative-Logarithmic Model Equivalence

How can the Derivative and Logarithmic model formulas, which look so different, produce similar results? Compare the Leverage Effect's relative (%) value creation formulas:

$$v_{LEV}^{Log.} \approx v_{LEV}^{Deriv.}$$

$$\frac{\ln(ER_2/ER_1)}{\ln(TEqV_2/TEqV_1)} \approx \frac{\Delta ER \cdot \overline{TEV}}{\Delta TEqV}$$

Continuous change (r_X) is $\ln(X_2/X_1)$ and standard change (R_X) is $(X_2 - X_1)/X_1$. When changes in X are small, generally within a few percent, it can be shown that $r_X \approx R_X$. However, when the denominator of R_X is replaced with the average value of X_1 and X_2, the values remain close over quite a wide range of X, so that:

$$\ln(X_2/X_1) \approx \frac{X_2 - X_1}{(X_1 + X_2)/2} = \frac{\Delta X}{\overline{X}}$$

Substituting this relationship into $v_{LEV}^{Log.}$ provides:

$$\frac{\ln(ER_2/ER_1)}{\ln(TEqV_2/TEqV_1)} \approx \frac{\Delta ER/\overline{ER}}{\Delta TEqV/\overline{TEqV}}$$

Equity ratio is defined as company equity value divided by enterprise valuation at time t. If the average values of ER, $TEqV$, and TEV have a similar relationship between times t_1 and t_2, then $\overline{ER}$ can be replaced with $\overline{TEqV}/\overline{TEV}$, which proves the original identity:

$$\frac{\ln(ER_2/ER_1)}{\ln(TEqV_2/TEqV_1)} \approx \frac{\Delta ER/\overline{ER}}{\Delta TEqV/\overline{TEqV}} = \frac{\Delta ER \cdot \dfrac{\overline{TEV}}{\overline{TEqV}}}{\Delta TEqV/\overline{TEqV}} = \frac{\Delta ER \cdot \overline{TEV}}{\Delta TEqV}$$

6. EBITDA and Valuation Multiples

This chapter adapts the Conventional Model's EBITDA Growth and Multiple Expansion to the superior leverage models described in Chapter 5. Company-level equity returns are expressed as the sum of EBITDA Growth (V_{EG}), Multiple Expansion (V_{ME}), and the Leverage Effect (V_{LEV}), where:

$$\Delta TEqV = V_{EG} + V_{ME} + V_{LEV}$$

$$= \Delta E \cdot \overline{M} \cdot \overline{ER} + \Delta M \cdot \overline{E} \cdot \overline{ER} + \Delta ER \cdot \overline{TEV}$$

$$= \Delta TEqV \frac{\ln(E_2/E_1)}{\ln(TEqV_2/TEqV_1)} + \Delta TEqV \frac{\ln(M_2/M_1)}{\ln(TEqV_2/TEqV_1)} + \Delta TEqV \frac{\ln(ER_2/ER_1)}{\ln(TEqV_2/TEqV_1)}$$

6.1. Background

In some interpretations of value creation analysis, GPs are given credit for EBITDA Growth, but not Multiple Expansion. Such models generally ascribe V_{EG} to operational impact or investment manager "skill", V_{ME} to market timing or "luck", and V_{LEV} to the impact of debt or "financial engineering".

The multitude of different value drivers introduced in Section 1.7 suggests that this is oversimplified. Manager-driven sales execution and operational improvements, as well as market-driven changes to raw material costs or addressable market size, impact EBITDA Growth. Market valuation trends, as well as company-specific size, growth rate, and earnings quality improvements, impact Multiple Expansion.

Each of these manager-driven and market-driven influences can be modeled as a distinct value driver. However, they are all downstream from V_{EG} and V_{ME} and, therefore, depend on how these two central value creation components are calculated. The Derivative and Logarithmic models are better foundations for such analyses because when the Conventional Model understates the debt component, it overstates V_{UNLEV}, V_{EG}, V_{ME}, and every other value driver that is a function of company valuations or P&Ls.

The Derivative and Logarithmic models appear to have little in common when comparing their derivations and value creation formulas. However, despite their differences, they provide nearly identical results with values generally within a few tenths of a percent, regardless of whether returns are expressed in the form of absolute ($) value creation, relative (%) value creation, equity return multiples or multipliers, or value creation IRRs.

6.2. Derivative Model

Chapter 4 demonstrated that company enterprise valuation (*TEV*) is the product of company EBITDA (*E*) and valuation multiple (*M*), so the change in company enterprise valuation can be expressed:

$$\Delta TEV = \Delta E \cdot \overline{M} + \Delta M \cdot \overline{E}$$

Additionally, Chapter 5 showed that company equity value (*TEqV*) is the product of *TEV* and equity ratio (*ER*), so company-level value creation can be expressed:

$$\Delta TEqV = \Delta TEV \cdot \overline{ER} + \Delta ER \cdot \overline{TEV}$$

Substituting the expression for ΔTEV into $\Delta TEqV$ provides:

$$\begin{aligned} \Delta TEqV &= \Delta TEV \cdot \overline{ER} + \Delta ER \cdot \overline{TEV} \\ &= (\Delta E \cdot \overline{M} + \Delta M \cdot \overline{E}) \cdot \overline{ER} + \Delta ER \cdot \overline{TEV} \\ &= \Delta E \cdot \overline{M} \cdot \overline{ER} + \Delta M \cdot \overline{E} \cdot \overline{ER} + \Delta ER \cdot \overline{TEV} \\ &= V_{EG} + V_{ME} + V_{LEV} \end{aligned}$$

The Derivative Model's $\Delta E \cdot \overline{M} \cdot \overline{ER}$ term is EBITDA Growth (V_{EG}), and its $\Delta M \cdot \overline{E} \cdot \overline{ER}$ term is Multiple Expansion (V_{ME}). These are equivalent to the Conventional Model formulas multiplied by the average percentage of equity in company capital structure, so that:

$$V_{EG}^{Deriv.} = V_{EG}^{Conv.} \cdot \overline{ER}$$
$$V_{ME}^{Deriv.} = V_{ME}^{Conv.} \cdot \overline{ER}$$

The relationship between the Derivative Model's Leverage Effect and the Conventional Model's Cash Flow Generation is described by the following identity:

$$V_{LEV}^{Deriv.} = V_{CF}^{Conv.} \cdot \frac{\Delta ER \cdot \overline{TEV}}{-\Delta ND}$$

Adding average enterprise valuation to AlphaCo's input variables provides:

Table 6.1. AlphaCo value creation input variables

	t_1	t_2	ΔX	$\overline{X}$
TEV	\$100	\$180	+\$80	\$140
TEqV	\$80	\$180	+\$100	-
ND	\$20	\$0	-\$20	-
ER	80%	100%	+20%	90%
E	\$20	\$30	+\$10	\$25
M	5.0x	6.0x	+1.0x	5.5x

Solving for AlphaCo's absolute (\$) value creation:

$$\Delta TEqV = V_{EG} + V_{ME} + V_{LEV}$$

$$= \Delta E \cdot \overline{M} \cdot \overline{ER} + \Delta M \cdot \overline{E} \cdot \overline{ER} + \Delta ER \cdot \overline{TEV}$$

$$= \$10 \cdot 5.5\text{x} \cdot 90\% + 1.0\text{x} \cdot \$25 \cdot 90\% + 20\% \cdot \$140$$

$$= \$49.5 + \$22.5 + \$28.0 = \$100.0$$

Relative (%) value creation can be determined from absolute (\$) value creation:

$$v_{EG} + v_{ME} + v_{LEV} = 100\%$$

$$\frac{V_{EG}}{\Delta TEqV} + \frac{V_{ME}}{\Delta TEqV} + \frac{V_{LEV}}{\Delta TEqV} = 100\%$$

$$\frac{\$49.5}{\$100} + \frac{\$22.5}{\$100} + \frac{\$28.0}{\$100} = 100\%$$

$$49.5\% + 22.5\% + 28.0\% = 100\%$$

The company equity return multiple (χ_{EQUITY}) is 2.25x, as determined by $TEqV_2/TEqV_1$. The equity return multiplier for each value driver is calculated:

$$\chi_{EQUITY} = \chi_{EG} \cdot \chi_{ME} \cdot \chi_{LEV}$$

$$= \chi_{EQUITY}{}^{v_{EG}} \cdot \chi_{EQUITY}{}^{v_{ME}} \cdot \chi_{EQUITY}{}^{v_{LEV}}$$

$$= 2.25x^{0.495} \cdot 2.25x^{0.225} \cdot 2.25x^{0.280}$$

$$= 1.49x \cdot 1.20x \cdot 1.25x = 2.25x$$

The company equity IRR (IRR_{EQUITY}) can be estimated from χ_{EQUITY} and the investment hold (Δt) in years:

$$IRR_{EQUITY} = \chi_{EQUITY}{}^{1/\Delta t} - 1$$
$$= 2.25x^{1/4.0} - 1 = 22.5\%$$

Value creation IRRs can be determined from equity return multipliers:

$$\left(1 + IRR_{EQUITY}\right) = (1 + IRR_{EG}) \cdot (1 + IRR_{ME}) \cdot (1 + IRR_{LEV})$$
$$\chi_{EQUITY}{}^{1/\Delta t} = \chi_{EG}{}^{1/\Delta t} \cdot \chi_{ME}{}^{1/\Delta t} \cdot \chi_{LEV}{}^{1/\Delta t}$$
$$2.25x^{1/4.0} = 1.49x^{1/4.0} \cdot 1.20x^{1/4.0} \cdot 1.25x^{1/4.0}$$
$$(1.225) = (1.106) \cdot (1.046) \cdot (1.058)$$
$$(1 + 22.5\%) = (1 + 10.6\%) \cdot (1 + 4.6\%) \cdot (1 + 5.8\%)$$

Thus, the Derivative Model provides a complete description of company-level EBITDA Growth, Multiple Expansion, and Leverage Effect.

6.3. Logarithmic Model

Chapter 5 demonstrated that $TEqV$ can be expressed as $TEV \cdot ER$, and Chapter 4 showed that TEV can be expressed as $E \cdot M$. Therefore:

$$TEqV = E \cdot M \cdot ER$$

The Logarithmic Model recognizes that if company equity value is the product of these components, changes in company equity value can be expressed as equity return multipliers:

$$\chi_{EQUITY} = \frac{TEqV_2}{TEqV_1} = \frac{(E_2 \cdot M_2 \cdot ER_2)}{(E_1 \cdot M_1 \cdot ER_1)} = \frac{E_2}{E_1} \cdot \frac{M_2}{M_1} \cdot \frac{ER_2}{ER_1}$$
$$= \chi_{EG} \cdot \chi_{ME} \cdot \chi_{LEV}$$

The company equity return multiple equals the product of the equity return multipliers for EBITDA Growth (χ_{EG}), Multiple Expansion (χ_{ME}), and Leverage Effect (χ_{LEV}).

AlphaCo's equity return multipliers can be calculated from the input variables above:

Table 6.2 AlphaCo value creation input variables

	t_1	t_2	X_2/X_1
TEV	\$100	\$180	1.80x
TEqV	\$80	\$180	2.25x
ND	\$20	\$0	n/a
ER	80%	100%	1.25x
E	\$20	\$30	1.50x
M	5.0x	6.0x	1.20x

$$\chi_{EQUITY} = \chi_{EG} \cdot \chi_{ME} \cdot \chi_{LEV}$$
$$= 1.50x \cdot 1.20x \cdot 1.25x = 2.25x$$

Value creation IRRs can be determined from equity return multipliers:

$$\left(1 + IRR_{EQUITY}\right) = (1 + IRR_{EG}) \cdot (1 + IRR_{ME}) \cdot (1 + IRR_{LEV})$$
$$\chi_{EQUITY}{}^{1/\Delta t} = \chi_{EG}{}^{1/\Delta t} \cdot \chi_{ME}{}^{1/\Delta t} \cdot \chi_{LEV}{}^{1/\Delta t}$$
$$2.25x^{1/4.0} = 1.50x^{1/4.0} \cdot 1.20x^{1/4.0} \cdot 1.25x^{1/4.0}$$
$$(1.225) = (1.107) \cdot (1.047) \cdot (1.057)$$
$$(1 + 22.5\%) = (1 + 10.7\%) \cdot (1 + 4.7\%) \cdot (1 + 5.7\%)$$

Measurements of relative (%) value creation and absolute (\$) value creation require a few more steps. The continuous change in company equity value between times t_1 and t_2 is:

$$r_{\text{EQUITY}} = \ln(TEqV_2) - \ln(TEqV_1) = \ln\left(\frac{TEqV_2}{TEqV_1}\right)$$

Since $TEqV_2/TEqV_1$ is χ_{EQUITY}, the continuous change is:

$$r_{\text{EQUITY}} = \ln\left(\chi_{EQUITY}\right) = \ln(\chi_{EG} \cdot \chi_{ME} \cdot \chi_{LEV})$$

Natural logarithms have the useful property that $\ln(a \cdot b \cdot c)$ equals the sum of $\ln(a)$, $\ln(b)$, and $\ln(c)$, so the continuous change in company equity value can be expressed:

$$r_{\text{EQUITY}} = \ln\left(\chi_{EQUITY}\right) = \ln(\chi_{EG}) + \ln(\chi_{ME}) + \ln(\chi_{LEV}) = r_{\text{EG}} + r_{\text{ME}} + r_{\text{LEV}}$$

The r_{EQUITY} expression suggests that the continuous change in company equity value is equal to the sum of the continuous changes in EBITDA, valuation multiple, and equity ratio. Dividing both sides by $\ln(\chi_{EQUITY})$ provides:

$$\frac{\ln(\chi_{EQUITY})}{\ln(\chi_{EQUITY})} = \frac{\ln(\chi_{EG})}{\ln(\chi_{EQUITY})} + \frac{\ln(\chi_{ME})}{\ln(\chi_{EQUITY})} + \frac{\ln(\chi_{LEV})}{\ln(\chi_{EQUITY})} = 100\%$$

$$v_{EG} + v_{ME} + v_{LEV} = 100\%$$

This is a useful way to measure company relative (%) value creation. It is not as intuitive as the Conventional or Derivative models, but it is often the fastest way to produce a result, especially when many value creation components are involved.

Solving for AlphaCo provides:

$$v_{EG} + v_{ME} + v_{LEV} = 100\%$$

$$\frac{\ln(1.50x)}{\ln(2.25x)} + \frac{\ln(1.20x)}{\ln(2.25x)} + \frac{\ln(1.25x)}{\ln(2.25x)} = 100\%$$

$$50.0\% + 22.5\% + 27.5\% = 100\%$$

Multiplying both sides of the relative (%) value creation formula by $\Delta TEqV$ provides a solution for absolute ($) value creation:

$$\begin{aligned} \Delta TEqV &= V_{EG} + V_{ME} + V_{LEV} \\ &= v_{EG} \cdot \Delta TEqV + v_{ME} \cdot \Delta TEqV + v_{LEV} \cdot \Delta TEqV \\ &= 50.0\% \cdot \$100 + 22.5\% \cdot \$100 + 27.5\% \cdot \$100 \\ &= \$50.0 + \$22.5 + \$27.5 = \$100.0 \end{aligned}$$

Thus, the Logarithmic Model provides a complete description of company-level EBITDA Growth, Multiple Expansion, and Leverage Effect, and generates results that are nearly identical to those provided by the Derivative Model.

6.4. Model Comparisons

The following table compares the return measurements and value drivers provided by the Conventional, Derivative, and Logarithmic models. Results are presented in the form of absolute ($) value creation, relative (%) value creation, equity return multiples or multipliers, and value creation IRRs.

Table 6.3 Comparison of AlphaCo's value drivers

	Conv.	Deriv.	Log.		Conv.	Deriv.	Log.
V_{EG}	*\$55.0*	*\$49.5*	*\$50.0*	x_{EG}	*1.56x*	*1.49x*	*1.50x*
V_{ME}	*\$25.0*	*\$22.5*	*\$22.5*	x_{ME}	*1.22x*	*1.20x*	*1.20x*
$V_{CF/LEV}$	*\$20.0*	*\$28.0*	*\$27.5*	$x_{CF/LEV}$	*1.18x*	*1.25x*	*1.25x*
V_{EQUITY}	*\$100.0*	*\$100.0*	*\$100.0*	x_{EQUITY}	*2.25x*	*2.25x*	*2.25x*
v_{EG}	*55.0%*	*49.5%*	*50.0%*	IRR_{EG}	*11.8%*	*10.6%*	*10.7%*
v_{ME}	*25.0%*	*22.5%*	*22.5%*	IRR_{ME}	*5.2%*	*4.6%*	*4.7%*
$v_{CF/LEV}$	*20.0%*	*28.0%*	*27.5%*	$IRR_{CF/LEV}$	*4.1%*	*5.8%*	*5.7%*
v_{EQUITY}	*100.0%*	*100.0%*	*100.0%*	IRR_{EQUITY}	*22.5%*	*22.5%*	*22.5%*

Table 6.3 validates several assertions:

- The Derivative and Logarithmic models are quite close.
- Conventional Model values generally understate the impact of leverage.
- After accounting for the different debt or leverage calculations, the relative ratio for every other value driver is about the same in all value creation models. For example, the ratio of V_{EG} to V_{ME} is precisely 2.20x in both the Conventional and Derivative models, and 2.22x in the Logarithmic Model.

Figure 6.1 compares AlphaCo's value creation IRRs provided by the Conventional, Derivative, and Logarithmic models:

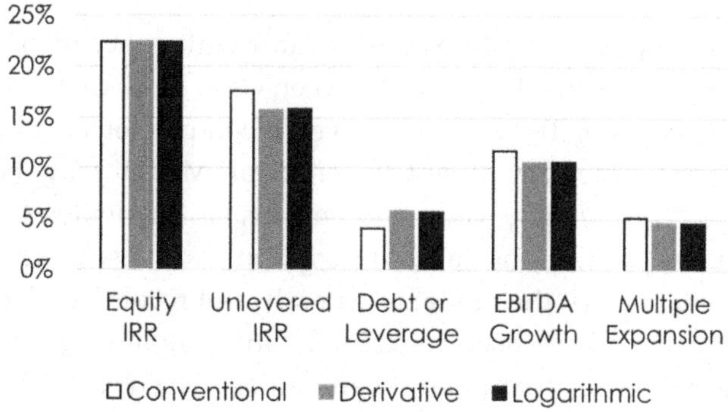

Figure 6.1 Value creation IRRs for AlphaCo

6.5. Summary

The three value creation models provide complete company-level descriptions of EBITDA Growth, Multiple Expansion, and the Leverage Effect or Cash Flow Generation. Each model can express returns and value drivers in the form of absolute ($) value creation, relative (%) value creation, equity return multiples or multipliers, and value creation IRRs.

The Derivative and Logarithmic models provide better representations of leverage. The Derivative Model is more intuitive and easier to apply when there are only a few value drivers, but the Logarithmic Model can provide answers with fewer input variables and fewer steps when there are several value creation components or when results are required in the form of equity return multipliers or value creation IRRs.

The Conventional Model tends to understate the impact of leverage and, therefore, overstate value creation attributed to the Unlevered Return, EBITDA Growth, and Multiple Expansion. The three models are related through the equity ratio (ER), where:

$$V_{UNLEV}^{Conv.} \cdot \overline{ER} = V_{UNLEV}^{Deriv.} \approx V_{UNLEV}^{Log.}$$

$$V_{EG}^{Conv.} \cdot \overline{ER} = V_{EG}^{Deriv.} \approx V_{EG}^{Log.}$$

$$V_{ME}^{Conv.} \cdot \overline{ER} = V_{ME}^{Deriv.} \approx V_{ME}^{Log.}$$

$$V_{CF}^{Conv.} \cdot \frac{\Delta ER \cdot \overline{TEV}}{-\Delta ND} = V_{LEV}^{Deriv.} \approx V_{LEV}^{Log.}$$

For a new analysis, it would be best to start with either the Derivative or Logarithmic model, as the difference between Cash Flow Generation and the Leverage Effect can be substantial. However, if comparisons to older analyses or data sets are necessary, the Conventional Model may be required. Consequently, *Private Equity Value Creation Analysis* provides Conventional Model formulas, derivations, and descriptions for every value creation component, even though the model is generally not recommended.

Thus far, analyses have been concerned with company equity values and the aggregate returns of all company shareholders. The next chapter examines return measurements and value drivers for a specific shareholder, such as a Fund or a GP.

6.6. Optional: Input Combination Errors

The Conventional and Derivative models are susceptible to minor errors when value creation input variables are processed in an order that differs from their derivations. Sometimes, numbers must be averaged and then multiplied. Other times, they must be multiplied and then averaged. For example, since company equity value, EBITDA, valuation multiple, and equity ratio are all functions of time, they may be written:

$$TEqV_{(t)} = E_{(t)} \cdot M_{(t)} \cdot ER_{(t)}$$

These terms can be expressed as time derivatives that are related by the "Product Rule":[4]

$$\frac{\partial TEqV_{(t)}}{\partial t} = \frac{\partial}{\partial t}\left(E_{(t)} \cdot M_{(t)} \cdot ER_{(t)}\right)$$
$$= \frac{\partial E_{(t)}}{\partial t} \cdot M_{(t)} \cdot ER_{(t)} + \frac{\partial M_{(t)}}{\partial t} \cdot E_{(t)} \cdot ER_{(t)} + \frac{\partial ER_{(t)}}{\partial t} \cdot E_{(t)} \cdot M_{(t)}$$

Replacing instantaneous changes (∂x) with discrete changes (Δx) and then multiplying both sides by Δt provides:

$$\Delta TEqV = \Delta E \cdot \overline{M} \cdot \overline{ER} + \Delta M \cdot \overline{E} \cdot \overline{ER} + \Delta ER \cdot \overline{E} \cdot \overline{M}$$
$$= V_{EG} + V_{ME} + V_{LEV}$$

Note the difference from the expression in Section 6.2, where V_{LEV} is rendered $\Delta ER \cdot \overline{TEV}$. Using the formula above to solve for absolute ($) value creation provides:

$$\Delta TEqV = V_{EG} + V_{ME} + V_{LEV}$$
$$= \Delta E \cdot \overline{M} \cdot \overline{ER} + \Delta M \cdot \overline{E} \cdot \overline{ER} + \Delta ER \cdot \overline{E} \cdot \overline{M}$$
$$= \$10 \cdot 5.5\text{x} \cdot 90\% + 1.0\text{x} \cdot \$25 \cdot 90\% + 20\% \cdot \$25 \cdot 5.5\text{x}$$
$$= \$49.5 + \$22.5 + \$27.5 = \$99.5$$

Company-level value creation is known to be $100.0, not $99.5. The error in V_{LEV} is relatively small, but it would flow through to v_{LEV}, χ_{LEV}, and IRR_{LEV}. It arises because $\overline{TEV}$ is $\overline{E \cdot M}$, which is not the same as $\overline{E} \cdot \overline{M}$. The former is the average of products, $100 and $180, which is $140. The latter is the product of averages, $25 and 5.5x, which is $137.5.

---

[4] A similar derivation, with additional steps and detail, is demonstrated in Section 4.2.

If this is a source of error, what about $\overline{M} \cdot \overline{ER}$ in V_{EG} and $\overline{E} \cdot \overline{ER}$ in V_{ME}? Replacing them with $\overline{M \cdot ER}$ and $\overline{E \cdot ER}$ would provide:

$$\Delta TEqV = V_{EG} + V_{ME} + V_{LEV}$$

$$= \Delta E \cdot \overline{M \cdot ER} + \Delta M \cdot \overline{E \cdot ER} + \Delta ER \cdot \overline{E \cdot M}$$

$$= \$10 \cdot 5.0\text{x} + 1.0\text{x} \cdot \$23 + 20\% \cdot \$140$$

$$= \$50.0 + \$23.0 + \$28.0 = \$101.0$$

In this case, company-level value creation is overstated by \$1.0. The term $\overline{M \cdot ER}$ corresponds to the t_1 and t_2 average $TEqV/E$, while the term $\overline{E \cdot ER}$ corresponds to the t_1 and t_2 average $TEqV/M$. They provide reasonable estimates of $\Delta TEqV$ because it is true that at any time t:

$$E_{(t)} \cdot M_{(t)} = TEqV_{(t)}/ER_{(t)}$$

$$E_{(t)} \cdot ER_{(t)} = TEqV_{(t)}/M_{(t)}$$

However, errors result because the products of averages are not the same as the averages of products. Leverage Effect is derived with $\overline{TEV}$, not $\overline{E} \cdot \overline{M}$. EBITDA Growth and Multiple Expansion are derived with $\overline{M}$ and $\overline{E}$, not $\overline{M \cdot ER}$ and $\overline{E \cdot ER}$. Such errors can generally be eliminated, or at least minimized, when value creation input variables preserve the basic structures that emerged from their derivations, as described in *Private Equity Value Creation Analysis*.

7. Fund Ownership and Isomeric Returns

This chapter describes value creation for a specific shareholder, like a Fund or GP. Prior chapters evaluated aggregate returns and value creation components for all company shareholders. Fund-level returns are expressed as the sum of EBITDA Growth (V_{EG}), Multiple Expansion (V_{ME}), Leverage Effect (V_{LEV}), and Fund Ownership Impact (V_{FO}), where:

$$\Delta FEqV = V_{EG} + V_{ME} + V_{LEV} + V_{FO}$$

$$= \Delta E \cdot \overline{M} \cdot \overline{ER} \cdot \overline{FO} + \Delta M \cdot \overline{E} \cdot \overline{ER} \cdot \overline{FO} + \Delta ER \cdot \overline{TEV} \cdot \overline{FO} + \Delta FO \cdot \overline{TEqV}$$

$$= \Delta FEqV \frac{\ln(E_2/E_1)}{\ln(MOIC)} + \Delta FEqV \frac{\ln(M_2/M_1)}{\ln(MOIC)} + \Delta TEqV \frac{\ln(ER_2/ER_1)}{\ln(MOIC)} + \Delta FEqV \frac{\ln(FO_2/FO_1)}{\ln(MOIC)}$$

7.1. Background

The company equity returns described thus far included the returns of the Fund or GP, as well as those of equity co-investors and company management. The latter's returns are important, but it is the Fund or GP returns that generate Carried Interest and drive LP portfolio alpha.

Unless otherwise noted, the following return measurements like MOIC always correspond to the Gross Returns for a specific Fund or GP. Value drivers such as EBITDA Growth and Multiple Expansion are also shareholder-specific, with notation like V_{EG} and V_{ME} unchanged from company-level value drivers described in prior chapters. Fund-level absolute ($) value creation ($V_{GROSS}$) between times t_1 and t_2, is the change in Fund equity value ($FEqV$), or $\Delta FEqV$:

$$V_{GROSS} = \Delta FEqV = FEqV_2 - FEqV_1$$

Fund-level equity return multiple (χ_{GROSS}), or Gross MOIC, is defined:

$$MOIC = \chi_{GROSS} = FEqV_2/FEqV_1$$

In contrast, company-level absolute ($) value creation is defined as $\Delta TEqV$ or V_{EQUITY}, and company-level equity return multiples are expressed as $TEqV_2/TEqV_1$ or χ_{EQUITY}.

7.2. A Shortcut: The J-Factor Method

One way to translate company-level value drivers into Fund-level value drivers is to use the *J*-Factor, defined as:

$$J = \Delta FEqV / \Delta TEqV$$

Since $\Delta FEqV$ is the product of the *J*-Factor and $\Delta TEqV$, Fund-level value creation can be expressed:

$$\begin{aligned} \Delta FEqV &= J \cdot \Delta TEqV \\ &= J \cdot (\Delta E \cdot \overline{M} \cdot \overline{ER} + \Delta M \cdot \overline{E} \cdot \overline{ER} + \Delta ER \cdot \overline{TEV}) \\ &= J \cdot \Delta E \cdot \overline{M} \cdot \overline{ER} + J \cdot \Delta M \cdot \overline{E} \cdot \overline{ER} + J \cdot \Delta ER \cdot \overline{TEV} \\ &= V_{EG} + V_{ME} + V_{LEV} \end{aligned}$$

This defines Fund-level value creation as the product of a *J*-Factor and company-level value creation. If an investor in AlphaCo generates a $\Delta FEqV$ of $44.0, then its *J*-Factor is $44.0/$100.0, or 44.0%, and:

$$\begin{aligned} \Delta FEqV &= V_{EG} + V_{ME} + V_{LEV} \\ &= 44.0\% \cdot \$49.5 + 44.0\% \cdot \$22.5 + 44.0\% \cdot \$28.0 \\ &= \$21.8 + \$9.9 + \$12.3 = \$44.0 \end{aligned}$$

The *J*-Factor approach effectively scales company-level value creation components to Fund-level value creation. It may be used when the Fund's ownership of company equity changes by no more than a few percent over the hold. The *J*-Factor makes no reference to specific company securities, which is a helpful simplification because ownership percentages can be complex and are often a source of errors when used as value creation input variables.

7.3. Fund Ownership

GPs often own several different securities for each investment: common stock, preferred stock, participating preferred stock, etc. The securities have complex interdependent relationships and values that fluctuate with time and circumstance. When an analyst asks a GP or deal team for an "ownership" number, there is no telling what they will receive. Different percentages can be correct, according to different definitions. Consequently, the ownership values provided in Fund reporting and GP marketing materials are often calculated inconsistently and are occasionally wrong.

To avoid ownership-driven errors, it is best to calculate Fund ownership (*FO*) directly from *FEqV* and *TEqV* values that are known to be accurate, where:

$$FO = FEqV/TEqV$$

Like the *J*-Factor, *FO* makes no reference to specific company securities. During the hold, *FO* is simply the fair value of company equity held by the Fund divided by equity held by all shareholders. At entry, *FO* is the Fund's invested equity capital divided by capital invested by all company shareholders. At exit, *FO* is proceeds to the Fund divided by proceeds to all company shareholders.

An input variable for *FO* is needed because the *J*-Factor approach begins to fail when *FO* changes by more than a few percent. The following example is extreme but illustrative. Consider an investment with relatively flat company-level equity value between times t_1 and t_2, where $\Delta TEqV$ is -\$0.1. However, the Fund owns participating preferred securities that change *FO*, causing $\Delta FEqV$ to increase by \$15.0 over the same period. The resulting *J*-Factor is -\$15/\$0.1, or -15,000%. In such a case, modest company-level EBITDA Growth of +\$2.0 would be characterized as Fund-level EBITDA Growth of -\$300.

This result is clearly spurious. Most of $\Delta FEqV$ is driven by changes in *FO*, not by changes in *E*, *M*, or *ER*. Such an investment requires a value creation model that accounts for ownership changes, perhaps by including some function of ΔFO as a distinct value creation component.

7.4. Conventional and Derivative Models

Fund-level equity value is the product of company-level equity value and Fund ownership. Therefore, changes in *FEqV* between times t_1 and t_2 can be expressed:

$$\begin{aligned}\Delta FEqV &= FEqV_2 - FEqV_1 \\ &= FO_2 \cdot TEqV_2 - FO_1 \cdot TEqV_1\end{aligned}$$

FEqV, *FO*, and *TEqV* are all functions of time, so at any time t:

$$FEqV_{(t)} = FO_{(t)} \cdot TEqV_{(t)}$$

These terms can be expressed as time derivatives that are related by the "Product Rule":[5]

$$\frac{\partial FEqV_{(t)}}{\partial t} = \frac{\partial}{\partial t}\left(FO_{(t)} \cdot TEqV_{(t)}\right) = \frac{\partial FO_{(t)}}{\partial t} \cdot TEqV_{(t)} + \frac{\partial TEqV_{(t)}}{\partial t} \cdot FO_{(t)}$$

Replacing instantaneous changes (∂x) with discrete changes (Δx) and then multiplying both sides by Δt provides:

$$\begin{aligned}\Delta FEqV &= \Delta FO \cdot \overline{TEqV} + \Delta TEqV \cdot \overline{FO} \\ &= V_{FO} + V_{ISO}\end{aligned}$$

The $\Delta FO \cdot \overline{TEqV}$ term is Fund Ownership Impact (V_{FO}). Negative numbers indicate dilution, perhaps from an expanding management option pool. Certain ownership structures, like participating preferred securities, can increase a Fund's share of company-level equity and generate positive values. The $\Delta TEqV \cdot \overline{FO}$ term is the equity return for all shareholders scaled to the Fund's investment size. It excludes Fund Ownership Impact, which can be different for different shareholders. It is defined as the Isomeric Return (V_{ISO}). Isomeric is from the Greek ισο-μερος, meaning "equal-share".

Consider a scenario where the lead investor in AlphaCo makes an initial investment of \$44.0, resulting in FO_1 of 55.0% (\$44/\$80). At exit, the Fund receives proceeds of \$88.0, resulting in FO_2 of 48.9% (\$88/\$180).

Table 7.1. AlphaCo value creation input variables

	t_1	t_2	ΔX	$\overline{X}$
TEV	\$100	\$180	+\$80	\$140
TEqV	\$80	\$180	+\$100	\$130
ND	\$20	\$0	-\$20	-
ER	80%	100%	+20%	90%
E	\$20	\$30	+\$10	\$25
M	5.0x	6.0x	+1.0x	5.5x
FEqV	\$44	\$88	+\$44	-
FO	55.0%	48.9%	-6.1%	51.9%

[5] A similar derivation, with additional steps and detail, is demonstrated in Section 4.2.

Fund-level value creation is clearly lagging company-level value creation because $FEqV_2/FEqV_1$ is 2.00x, while $TEqV_2/TEqV_1$ is 2.25x. This is indicative of dilution, which is common in successful investments where company management teams can earn incentive equity by achieving pre-determined growth initiatives. Increasing management team ownership lowers the Fund's percentage ownership of $TEqV$, which is manifested above as a negative ΔFO value.

AlphaCo's absolute ($) value creation formula provides:

$$\Delta FEqV = V_{FO} + V_{ISO}$$
$$= \Delta FO \cdot \overline{TEqV} + \Delta TEqV \cdot \overline{FO}$$
$$= -6.1\% \cdot \$130 + \$100 \cdot 51.9\%$$
$$= -\$7.9 + \$51.9 = \$44.0$$

The Isomeric Return of $51.9 is the hypothetical value creation if the Fund maintains 51.9% ownership of $TEqV$ throughout the hold. Because of dilution, the Fund only generates $44.0. The difference, -$7.9, is Fund Ownership Impact caused by the decrease in FO from 55.0% to 48.9%.

This framework of Isomeric Return and Fund Ownership Impact can be applied to the Conventional, Derivative, and Logarithmic models of value creation to provide Fund-level value creation components.

Conventional Model

Having established that company-level value creation can be expressed:

$$\Delta TEqV = V_{EG} + V_{ME} + V_{CF}$$
$$= \Delta E \cdot \overline{M} + \Delta M \cdot \overline{E} - \Delta ND$$

The $\Delta TEqV$-term can be substituted into the Fund-level value creation formula, providing:

$$\Delta FEqV = \Delta TEqV \cdot \overline{FO} + \Delta FO \cdot \overline{TEqV}$$
$$= (\Delta E \cdot \overline{M} + \Delta M \cdot \overline{E} - \Delta ND) \cdot \overline{FO} + \Delta FO \cdot \overline{TEqV}$$
$$= \Delta E \cdot \overline{M} \cdot \overline{FO} + \Delta M \cdot \overline{E} \cdot \overline{FO} - \Delta ND \cdot \overline{FO} + \Delta FO \cdot \overline{TEqV}$$
$$= V_{EG} + V_{ME} + V_{CF} + V_{FO}$$

Solving for AlphaCo provides:

$$\Delta FEqV = V_{EG} + V_{ME} + V_{CF} + V_{FO}$$
$$= \$10 \cdot 5.5\text{x} \cdot 51.9\% + 1.0\text{x} \cdot \$25 \cdot 51.9\% + \$20 \cdot 51.9\% - 6.1\% \cdot \$130$$
$$= \$28.6 + \$13.0 + \$10.4 - \$7.9 = \$44.0$$

The other measures of value creation are described as follows. Relative (%) value creation can be determined from absolute ($) value creation:

$$v_{EG} + v_{ME} + v_{CF} + v_{FO} = 100\%$$
$$\frac{V_{EG}}{\Delta FEqV} + \frac{V_{ME}}{\Delta FEqV} + \frac{V_{CF}}{\Delta FEqV} + \frac{V_{FO}}{\Delta FEqV} = 100\%$$

Equity return multipliers can be determined from relative (%) value creation:

$$MOIC = \chi_{EG} \cdot \chi_{ME} \cdot \chi_{CF} \cdot \chi_{FO}$$
$$MOIC = MOIC^{v_{EG}} \cdot MOIC^{v_{ME}} \cdot MOIC^{v_{CF}} \cdot MOIC^{v_{FO}}$$

Value creation IRRs can be determined from equity return multipliers:

$$(1 + IRR_{GROSS}) = (1 + IRR_{EG}) \cdot (1 + IRR_{ME}) \cdot (1 + IRR_{CF}) \cdot (1 + IRR_{FO})$$
$$MOIC^{1/\Delta t} = \chi_{EG}{}^{1/\Delta t} \cdot \chi_{ME}{}^{1/\Delta t} \cdot \chi_{CF}{}^{1/\Delta t} \cdot \chi_{FO}{}^{1/\Delta t}$$

Derivative Model

Having established that company-level value creation can be expressed:

$$\Delta TEqV = V_{EG} + V_{ME} + V_{LEV}$$
$$= \Delta E \cdot \overline{M} \cdot \overline{ER} + \Delta M \cdot \overline{E} \cdot \overline{ER} + \Delta ER \cdot \overline{TEV}$$

The $\Delta TEqV$-term can be substituted into $\Delta FEqV$, providing:

$$\Delta FEqV = \Delta TEqV \cdot \overline{FO} + \Delta FO \cdot \overline{TEqV}$$
$$= (\Delta E \cdot \overline{M} \cdot \overline{ER} + \Delta M \cdot \overline{E} \cdot \overline{ER} + \Delta ER \cdot \overline{TEV}) \cdot \overline{FO} + \Delta FO \cdot \overline{TEqV}$$
$$= \Delta E \cdot \overline{M} \cdot \overline{ER} \cdot \overline{FO} + \Delta M \cdot \overline{E} \cdot \overline{ER} \cdot \overline{FO} + \Delta ER \cdot \overline{TEV} \cdot \overline{FO} + \Delta FO \cdot \overline{TEqV}$$
$$= V_{EG} + V_{ME} + V_{LEV} + V_{FO}$$

Solving for AlphaCo provides:

$$\Delta FEqV = V_{EG} + V_{ME} + V_{LEV} + V_{FO}$$
$$= \$10 \cdot 5.5\text{x} \cdot 90\% \cdot 51.9\% + 1.0\text{x} \cdot \$25 \cdot 90\% \cdot 51.9\% + 20\% \cdot \$140 \cdot 51.9\% - 6.1\% \cdot \$130$$
$$= \$25.7 + \$11.7 + \$14.5 - \$7.9 = \$44.0$$

Compared to the *J*-Factor approach described in Section 7.2, the value drivers for EBITDA Growth, Multiple Expansion, and the Leverage Effect are each about 18% higher to account for dilution, which is evident by Fund Ownership Impact of -$7.9. The value drivers for V_{EG}, V_{ME}, and V_{LEV} sum to $51.9, which is the Fund investment's absolute ($) Isomeric Return.

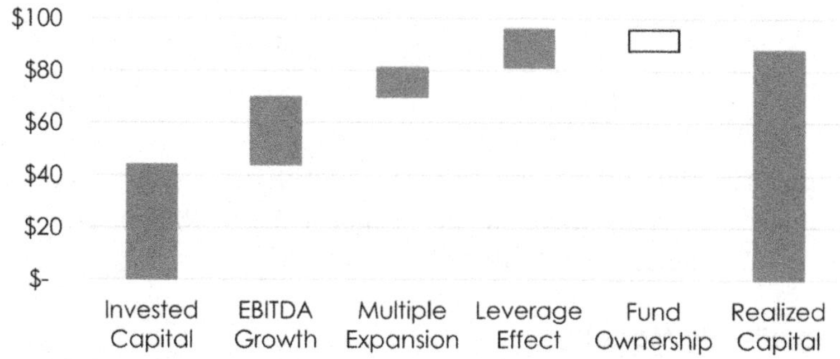

Figure 7.1 Derivative Model absolute ($) value creation bridge for AlphaCo

Relative (%) value creation can be determined from absolute ($) value creation:

$$v_{EG} + v_{ME} + v_{LEV} + v_{FO} = 100\%$$

$$\frac{V_{EG}}{\Delta FEqV} + \frac{V_{ME}}{\Delta FEqV} + \frac{V_{LEV}}{\Delta FEqV} + \frac{V_{FO}}{\Delta FEqV} = 100\%$$

Solving for AlphaCo provides:

$$v_{EG} + v_{ME} + v_{LEV} + v_{FO} = 100\%$$

$$\frac{\$25.7}{\$44.0} + \frac{\$11.7}{\$44.0} + \frac{\$14.5}{\$44.0} + \frac{-\$7.9}{\$44.0} = 100\%$$

$$58.4\% + 26.6\% + 33.1\% + -18.1\% = 100\%$$

The value drivers for v_{EG}, v_{ME}, and v_{LEV} sum to 118.1%, which is the Fund investment's relative (%) Isomeric Return (v_{ISO}).

Equity return multipliers can be determined from relative (%) value creation:

$$MOIC = \chi_{EG} \cdot \chi_{ME} \cdot \chi_{LEV} \cdot \chi_{FO}$$

$$= MOIC^{v_{EG}} \cdot MOIC^{v_{ME}} \cdot MOIC^{v_{LEV}} \cdot MOIC^{v_{FO}}$$

$$= 2.00x^{0.584} \cdot 2.00x^{0.266} \cdot 2.00x^{0.331} \cdot 2.00x^{-0.181}$$

$$= 1.50x \cdot 1.20x \cdot 1.26x \cdot 0.88x = 2.00x$$

The product of χ_{EG}, χ_{ME}, and χ_{LEV} values is 2.25x, which is the Fund investment's Isomeric Return multiple (χ_{ISO}). This value is identical to the company equity return multiple ($TEqV_2/TEqV_1$ or χ_{EQUITY}).

Value creation IRRs can be determined from equity return multipliers:

$$(1 + IRR_{GROSS}) = (1 + IRR_{EG}) \cdot (1 + IRR_{ME}) \cdot (1 + IRR_{LEV}) \cdot (1 + IRR_{FO})$$

$$MOIC^{1/\Delta t} = \chi_{EG}{}^{1/\Delta t} \cdot \chi_{ME}{}^{1/\Delta t} \cdot \chi_{LEV}{}^{1/\Delta t} \cdot \chi_{FO}{}^{1/\Delta t}$$

$$2.00x^{1/4.0} = 1.50x^{1/4.0} \cdot 1.20x^{1/4.0} \cdot 1.26x^{1/4.0} \cdot 0.88x^{1/4.0}$$

$$(1 + 18.9\%) = (1 + 10.7\%) \cdot (1 + 4.7\%) \cdot (1 + 5.9\%) \cdot (1 - 3.1\%)$$

The product of $(1 + IRR_{EG}) \cdot (1 + IRR_{ME}) \cdot (1 + IRR_{LEV})$ values less 100% is 22.5%, which is the Fund investment's Isomeric IRR (IRR_{ISO}). This value is identical to the company equity IRR (IRR_{EQUITY}).

7.5. Logarithmic Model

The Logarithmic Model recognizes that if Fund equity value is the product of company equity value and Fund ownership, and if company equity value is the product of EBITDA, valuation multiple, and equity ratio, then they can be incorporated into the equity return multiplier expression for the Fund investment's Gross MOIC:

$$MOIC = \frac{FEqV_2}{FEqV_1} = \frac{TEqV_2 \cdot FO_2}{TEqV_1 \cdot FO_1} = \frac{(E_2 \cdot M_2 \cdot ER_2) \cdot FO_2}{(E_1 \cdot M_1 \cdot ER_1) \cdot FO_1}$$

$$= \frac{TEqV_2}{TEqV_1} \cdot \frac{FO_2}{FO_1} = \frac{E_2}{E_1} \cdot \frac{M_2}{M_1} \cdot \frac{ER_2}{ER_1} \cdot \frac{FO_2}{FO_1}$$

$$\chi_{GROSS} = \chi_{EQUITY} \cdot \chi_{FO} = \chi_{EG} \cdot \chi_{ME} \cdot \chi_{LEV} \cdot \chi_{FO}$$

The MOIC and χ_{GROSS} above are identical to the Gross MOIC commonly found in Fund reporting and GP marketing materials, and they are equal to the product of the equity return multipliers for EBITDA Growth (χ_{EG}), Multiple Expansion (χ_{ME}), Leverage Effect (χ_{LEV}), and Fund Ownership Impact (χ_{FO}).

AlphaCo's equity return multiples and multipliers can be calculated from the input variables above:

Table 7.2 AlphaCo value creation input variables

	t_1	t_2	X_2/X_1	Component
TEV	\$100	\$180	1.80x	χ_{UNLEV}
TEqV	\$80	\$180	2.25x	χ_{EQUITY} or χ_{ISO}
ND	\$20	\$0	n/a	n/a
ER	80%	100%	1.25x	χ_{LEV}
E	\$20	\$30	1.50x	χ_{EG}
M	5.0x	6.0x	1.20x	χ_{ME}
FEqV	\$44	\$88	2.00x	*MOIC* or χ_{GROSS}
FO	55.0%	48.9%	0.89x	χ_{FO}

$$MOIC = \chi_{EG} \cdot \chi_{ME} \cdot \chi_{LEV} \cdot \chi_{FO}$$
$$= 1.50\text{x} \cdot 1.20\text{x} \cdot 1.25\text{x} \cdot 0.89\text{x} = 2.00\text{x}$$

The relative performance of χ_{GROSS} and χ_{EQUITY} is driven by χ_{FO}. Whenever there is dilution, ΔFO is negative and the χ_{FO}-term is less than 1.0x. This lowers the Fund investment's Gross MOIC relative to the equity return multiple for all company shareholders.

The other measures of value creation are described as follows. Value creation IRRs can be determined from equity return multipliers:

$$(1 + IRR_{GROSS}) = (1 + IRR_{EG}) \cdot (1 + IRR_{ME}) \cdot (1 + IRR_{LEV}) \cdot (1 + IRR_{FO})$$
$$MOIC^{1/\Delta t} = \chi_{EG}{}^{1/\Delta t} \cdot \chi_{ME}{}^{1/\Delta t} \cdot \chi_{LEV}{}^{1/\Delta t} \cdot \chi_{FO}{}^{1/\Delta t}$$
$$2.00x^{1/4.0} = 1.50x^{1/4.0} \cdot 1.20x^{1/4.0} \cdot 1.25x^{1/4.0} \cdot 0.89x^{1/4.0}$$
$$(1.189) = (1.107) \cdot (1.047) \cdot (1.057) \cdot (0.971)$$
$$(1 + 18.9\%) = (1 + 10.7\%) \cdot (1 + 4.7\%) \cdot (1 + 5.7\%) \cdot (1 - 2.9\%)$$

Relative (%) value creation can be determined from equity return multipliers:

$$v_{EG} + v_{ME} + v_{LEV} + v_{FO} = 100\%$$
$$\frac{\ln(\chi_{EG})}{\ln(MOIC)} + \frac{\ln(\chi_{ME})}{\ln(MOIC)} + \frac{\ln(\chi_{LEV})}{\ln(MOIC)} + \frac{\ln(\chi_{FO})}{\ln(MOIC)} = 100\%$$
$$\frac{\ln(1.50x)}{\ln(2.00x)} + \frac{\ln(1.20x)}{\ln(2.00x)} + \frac{\ln(1.25x)}{\ln(2.00x)} + \frac{\ln(0.89x)}{\ln(2.00x)} = 100\%$$
$$58.5\% + 26.3\% + 32.2\% - 17.0\% = 100\%$$

Relative (%) value creation is additive, so the sum of v_{EG} and v_{ME} is the Unlevered Return (v_{UNLEV}) of 84.8%, the sum of v_{UNLEV} and v_{LEV} is the Isomeric Return (v_{ISO}) of 117.0%, and the sum of v_{ISO} and v_{FO} corresponds to Fund-level value creation, which in relative (%) terms is always 100%. These relationships can be visualized with a stacked horizontal bar chart.

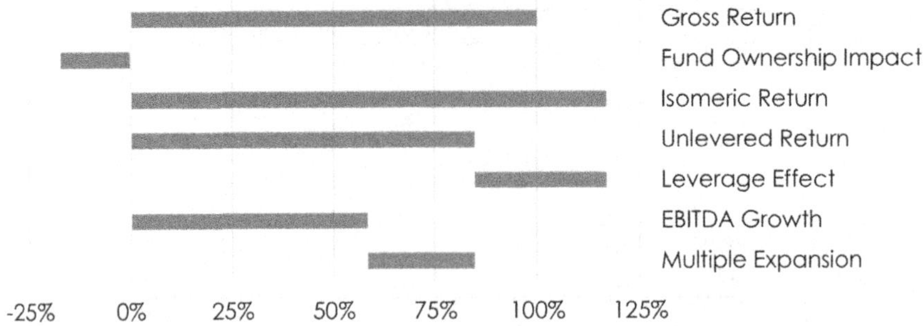

Figure 7.2 Logarithmic Model relative (%) value creation for AlphaCo

Absolute ($) value creation can be determined from relative (%) value creation:

$$\Delta FEqV = V_{EG} + V_{ME} + V_{LEV} + V_{FO}$$
$$= v_{EG} \cdot \Delta FEqV + v_{ME} \cdot \Delta FEqV + v_{LEV} \cdot \Delta FEqV + v_{FO} \cdot \Delta FEqV$$
$$= 58.5\% \cdot \$44 + 26.3\% \cdot \$44 + 32.2\% \cdot \$44 - 17.0\% \cdot \$44$$
$$= \$25.7 + \$11.6 + \$14.2 - \$7.5 = \$44.0$$

7.6. Model Comparisons

The following table compares the return measurements and value drivers provided by the Conventional, Derivative, and Logarithmic models. Results are presented in the form of absolute ($) value creation, relative (%) value creation, equity return multiples or multipliers, and value creation IRRs.

Table 7.3 Comparison of AlphaCo's value drivers

	Conv.	Deriv.	Log.		Conv.	Deriv.	Log.
V_{EG}	\$28.6	\$25.7	\$25.7	x_{EG}	1.57x	1.50x	1.50x
V_{ME}	\$13.0	\$11.7	\$11.6	x_{ME}	1.23x	1.20x	1.20x
$V_{CF/LEV}$	\$10.4	\$14.5	\$14.2	$x_{CF/LEV}$	1.18x	1.26x	1.25x
V_{FO}	-\$7.9	-\$7.9	-\$7.5	x_{FO}	0.88x	0.88x	0.89x
V_{GROSS}	\$44.0	\$44.0	\$44.0	x_{GROSS}	2.00x	2.00x	2.00x
v_{EG}	64.9%	58.4%	58.5%	IRR_{EG}	11.9%	10.7%	10.7%
v_{ME}	29.5%	26.6%	26.3%	IRR_{ME}	5.2%	4.7%	4.7%
$v_{CF/LEV}$	23.6%	33.1%	32.2%	$IRR_{CF/LEV}$	4.2%	5.9%	5.7%
v_{FO}	-18.1%	-18.1%	-17.0%	IRR_{FO}	-3.1%	-3.1%	-2.9%
v_{GROSS}	100%	100%	100%	IRR_{GROSS}	18.9%	18.9%	18.9%

The Conventional and Derivative model values for Fund Ownership Impact are identical because their formulas are identical. The same is true of

Isomeric Returns, Net Returns, and the value creation components for Fees & Expenses and Carried Interest described in Chapter 14. The Conventional and Derivative models start to diverge only after debt begins to influence the value drivers, with the incorporation of $\overline{ER}$. This is the case with the Unlevered Return, EBITDA Growth, Multiple Expansion, and the value creation components described in Chapter 8 through Chapter 12.

7.7. Summary

The three value creation models provide complete Fund-level descriptions of EBITDA Growth, Multiple Expansion, Leverage Effect or Cash Flow Generation, and Fund Ownership Impact. Each model can express returns and value drivers in the form of absolute ($) value creation, relative (%) value creation, equity return multiples or multipliers, and value creation IRRs.

Fund Ownership Impact can be substantial, and new value creation analyses should generally include it as a distinct value driver. If analysts must compare results to older analyses or data sets that exclude Fund Ownership Impact, then the *J*-Factor approach described in Section 7.2 may be used. However, analysts should be aware that the *J*-Factor-driven results can be skewed and even meaningless for investments that have significant Fund Ownership changes occurring over the hold.

The next chapter considers the impact that changes in company P&Ls, specifically revenue and EBITDA margin changes, have on value creation analysis.

8. Revenue and EBITDA Margins

This chapter describes EBITDA Growth in terms of its underlying components of Revenue Growth and EBITDA Margin Expansion. Fund-level returns are expressed as the sum of Revenue Growth (V_{RG}), EBITDA Margin Expansion (V_{EME}), Multiple Expansion (V_{ME}), Leverage Effect (V_{LEV}), and Fund Ownership Impact (V_{FO}), where:

$$\Delta FEqV = V_{RG} + V_{EME} + V_{ME} + V_{LEV} + V_{FO}$$

$$= \Delta R \cdot \overline{EM} \cdot \overline{M} \cdot \overline{ER} \cdot \overline{FO} + \Delta EM \cdot \overline{R} \cdot \overline{M} \cdot \overline{ER} \cdot \overline{FO} + \Delta M \cdot \overline{E} \cdot \overline{ER} \cdot \overline{FO} + \Delta ER \cdot \overline{TEV} \cdot \overline{FO} + \Delta FO \cdot \overline{TEqV}$$

$$= \Delta FEqV \frac{\ln(R_2/R_1)}{\ln(MOIC)} + \Delta FEqV \frac{\ln(EM_2/EM_1)}{\ln(MOIC)} + \Delta FEqV \frac{\ln(M_2/M_1)}{\ln(MOIC)} + \Delta FEqV \frac{\ln(ER_2/ER_1)}{\ln(MOIC)} + \Delta FEqV \frac{\ln(FO_2/FO_1)}{\ln(MOIC)}$$

8.1. Background

Revenue Growth is an obvious candidate for a value creation component because almost any discussion of private company performance includes revenue (R). Enterprise valuations are sometimes expressed as multiples of revenue, especially in early stage, growth, or distressed investments with negligible or negative EBITDA (E) values.

For such companies, R simply replaces E in every value creation formula described in the previous chapters. The valuation multiple (M) becomes TEV/R, ΔE becomes ΔR, $\overline{E}$ becomes $\overline{R}$, etc. The EBITDA Growth expressions become Revenue Growth expressions that are abbreviated V_{RG}, v_{RG}, χ_{RG}, and IRR_{RG}, as in Table 8.1.

Table 8.1 Revenue Growth formulas for revenue-based enterprise valuations

	Conventional Model	Derivative Model	Logarithmic Model
V_{RG}	$\Delta R \cdot \overline{M} \cdot \overline{FO}$	$\Delta R \cdot \overline{M} \cdot \overline{ER} \cdot \overline{FO}$	$\Delta FEqV \dfrac{\ln(R_2/R_1)}{\ln(MOIC)}$
v_{RG}	$\dfrac{\Delta R \cdot \overline{M} \cdot \overline{FO}}{\Delta FEqV}$	$\dfrac{\Delta R \cdot \overline{M} \cdot \overline{ER} \cdot \overline{FO}}{\Delta FEqV}$	$\dfrac{\ln(R_2/R_1)}{\ln(MOIC)}$
x_{RG}	$MOIC^{\frac{\Delta R \cdot \overline{M} \cdot \overline{FO}}{\Delta FEqV}}$	$MOIC^{\frac{\Delta R \cdot \overline{M} \cdot \overline{ER} \cdot \overline{FO}}{\Delta FEqV}}$	R_2/R_1
IRR_{RG}	$MOIC^{\frac{\Delta R \cdot \overline{M} \cdot \overline{FO}}{\Delta FEqV \cdot \Delta t}} - 1$	$MOIC^{\frac{\Delta R \cdot \overline{M} \cdot \overline{ER} \cdot \overline{FO}}{\Delta FEqV \cdot \Delta t}} - 1$	$(R_2/R_1)^{1/\Delta t} - 1$

For companies with EBITDA-based enterprise valuations, Revenue Growth emerges as a component of EBITDA Growth and is accompanied by a driver for EBITDA Margin Expansion. Since EBITDA margin (EM) is EBITDA divided by revenue, multiplying E by 1.0x in the form of R/R provides:

$$E = E \cdot \left(\frac{R}{R}\right) = R \cdot \left(\frac{E}{R}\right) = R \cdot EM$$

Improvements in either R or EM must enhance shareholder value. This results from the fact that Fund equity value is equal to $FO \cdot TEqV$, and company equity value is equal to $TEV \cdot ER$, and company enterprise valuation is equal to $E \cdot M$, and EBITDA is equal to $R \cdot EM$. Therefore, any increase in either R or EM must also increase $FEqV$.

The value creation relationships for Revenue Growth and EBITDA Margin Expansion are intuitive and straight forward. They also benefit from relying on value creation input variables that are widely available. Fund reporting and GP marketing materials often include summary versions of company P&Ls. Therefore, the required E, R, and EM values can typically be obtained without making special requests to GPs or deal teams.

8.2. Conventional and Derivative Models

Company EBITDA can be expressed as the product of revenue and EBITDA margin. Changes in E between times t_1 and t_2 are, therefore:

$$\begin{aligned} \Delta E &= E_2 - E_1 \\ &= R_2 \cdot EM_2 - R_1 \cdot EM_1 \end{aligned}$$

E, R, and EM are all functions of time, so at any time t:

$$E_{(t)} = R_{(t)} \cdot EM_{(t)}$$

These terms can be expressed as time derivatives that are related by the "Product Rule":[6]

$$\frac{\partial E_{(t)}}{\partial t} = \frac{\partial}{\partial t}\left(R_{(t)} \cdot EM_{(t)}\right) = \frac{\partial R_{(t)}}{\partial t} \cdot EM_{(t)} + \frac{\partial EM_{(t)}}{\partial t} \cdot R_{(t)}$$

Replacing instantaneous changes (∂x) with discrete changes (Δx) and then multiplying both sides by Δt provides:

$$\Delta E = \Delta R \cdot \overline{EM} + \Delta EM \cdot \overline{R}$$

Consider a scenario where AlphaCo increases revenue from \$80 to \$100 as it increases EBITDA from \$20 to \$30. In this case, EBITDA margin increases from EM_1 of 25% (\$20/\$80) to EM_2 of 30% (\$30/\$100).

Table 8.2. AlphaCo value creation input variables

	t_1	t_2	ΔX	$\overline{X}$
TEV	\$100	\$180	+\$80	\$140
TEqV	\$80	\$180	+\$100	\$130
ND	\$20	\$0	-\$20	-
ER	80%	100%	+20%	90%
E	\$20	\$30	+\$10	\$25
M	5.0x	6.0x	+1.0x	5.5x
R	\$80	\$100	+\$20	\$90
EM	25%	30%	+5%	27.5%
FEqV	\$44	\$88	+44	-
FO	55.0%	48.9%	-6.1%	51.9%

Solving the formula for ΔE provides:

$$\begin{aligned}\Delta E &= \Delta R \cdot \overline{EM} + \Delta EM \cdot \overline{R} \\ &= \$20 \cdot 27.5\% + 5.0\% \cdot \$90 \\ &= \$5.5 + \$4.5 = \$10.0\end{aligned}$$

Since the expression $\Delta R \cdot \overline{EM} + \Delta EM \cdot \overline{R}$ is identical to $E_2 - E_1$, it can be substituted for ΔE in the Conventional and Derivative model formulas for a Fund investment's EBITDA Growth (V_{EG}). The substitution creates two new value drivers: a term including $\Delta R \cdot \overline{EM}$ that represents Revenue Growth

[6] A similar derivation, with additional steps and detail, is demonstrated in Section 4.2.

(V_{RG}) and a term including $\Delta EM \cdot \overline{R}$ that represents EBITDA Margin Expansion (V_{EME}).

Conventional Model

Having established that Fund-level EBITDA Growth (V_{EG}) can be expressed:

$$V_{EG} = \Delta E \cdot \overline{M} \cdot \overline{FO}$$

Replacing ΔE with the expression above provides absolute ($) value creation:

$$V_{EG} = (\Delta R \cdot \overline{EM} + \Delta EM \cdot \overline{R}) \cdot \overline{M} \cdot \overline{FO}$$
$$= \Delta R \cdot \overline{EM} \cdot \overline{M} \cdot \overline{FO} + \Delta EM \cdot \overline{R} \cdot \overline{M} \cdot \overline{FO}$$
$$= V_{RG} + V_{EME}$$

Relative (%) value creation can be determined from absolute ($) value creation:

$$v_{EG} = v_{RG} + v_{EME}$$
$$\frac{V_{EG}}{\Delta FEqV} = \frac{V_{RG}}{\Delta FEqV} + \frac{V_{EME}}{\Delta FEqV}$$

Equity return multipliers can be determined from relative (%) value creation:

$$x_{EG} = x_{RG} \cdot x_{EME}$$
$$MOIC^{v_{EG}} = MOIC^{v_{RG}} \cdot MOIC^{v_{EME}}$$

Value creation IRRs can be determined from equity return multipliers:

$$(1 + IRR_{EG}) = (1 + IRR_{RG}) \cdot (1 + IRR_{EME})$$
$${x_{EG}}^{1/\Delta t} = {x_{RG}}^{1/\Delta t} \cdot {x_{EME}}^{1/\Delta t}$$

Replacing EBITDA Growth with Revenue Growth and EBITDA Margin Expansion has no impact on the other value drivers, which are calculated as described in earlier chapters. Compared to the Revenue Growth formulas for companies with revenue-based valuations (Table 8.1), the formulas for companies with EBITDA-based valuations include an additional term, $\overline{EM}$, for the t_1 and t_2 average EBITDA margin.

Derivative Model

Having established that Fund-level EBITDA Growth (V_{EG}) can be expressed:

$$V_{EG} = \Delta E \cdot \overline{M} \cdot \overline{ER} \cdot \overline{FO}$$

Replacing ΔE with the expression above provides absolute ($) value creation:

$$V_{EG} = (\Delta R \cdot \overline{EM} + \Delta EM \cdot \overline{R}) \cdot \overline{M} \cdot \overline{ER} \cdot \overline{FO}$$
$$= \Delta R \cdot \overline{EM} \cdot \overline{M} \cdot \overline{ER} \cdot \overline{FO} + \Delta EM \cdot \overline{R} \cdot \overline{M} \cdot \overline{ER} \cdot \overline{FO}$$
$$= V_{RG} + V_{EME}$$

Solving for AlphaCo provides:

$$V_{EG} = V_{RG} + V_{EME}$$
$$= \$20 \cdot 27.5\% \cdot 5.5x \cdot 90\% \cdot 51.9\% + 5\% \cdot \$90 \cdot 5.5x \cdot 90\% \cdot 51.9\%$$
$$= \$14.1 + \$11.6 = \$25.7$$

The value creation components V_{RG} and V_{EME} can replace V_{EG} in the expression for $\Delta FEqV$ described in Chapter 7, providing:

$$\Delta FEqV = V_{EG} + V_{ME} + V_{LEV} + V_{FO}$$
$$= V_{RG} + V_{EME} + V_{ME} + V_{LEV} + V_{FO}$$

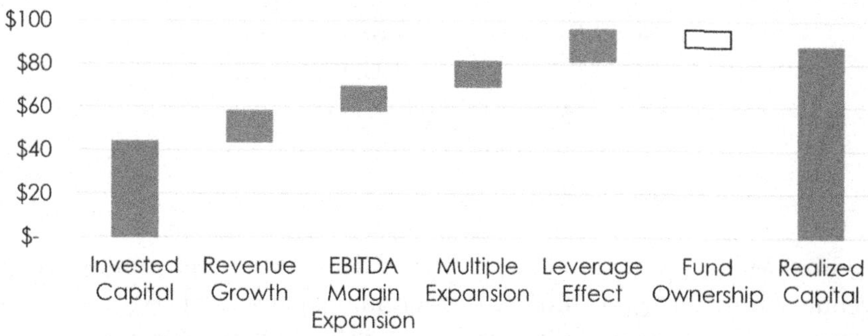

Figure 8.1 Derivative Model absolute ($) value creation bridge for AlphaCo

Relative (%) value creation can be determined from absolute ($) value creation:

$$v_{EG} = v_{RG} + v_{EME}$$
$$\frac{V_{EG}}{\Delta FEqV} = \frac{V_{RG}}{\Delta FEqV} + \frac{V_{EME}}{\Delta FEqV}$$

Equity return multipliers can be determined from relative (%) value creation:

$$x_{EG} = x_{RG} \cdot x_{EME}$$
$$MOIC^{v_{EG}} = MOIC^{v_{RG}} \cdot MOIC^{v_{EME}}$$

Value creation IRRs can be determined from equity return multipliers:

$$(1 + IRR_{EG}) = (1 + IRR_{RG}) \cdot (1 + IRR_{EME})$$

$$\chi_{EG}{}^{1/\Delta t} = \chi_{RG}{}^{1/\Delta t} \cdot \chi_{EME}{}^{1/\Delta t}$$

Replacing EBITDA Growth with Revenue Growth and EBITDA Margin Expansion has no impact on the other value drivers, which are calculated as described in earlier chapters. Compared to the Revenue Growth formulas for companies with revenue-based valuations (Table 8.1), the formulas for companies with EBITDA-based valuations include an additional term, $\overline{EM}$, for the t_1 and t_2 average EBITDA margin.

8.3. Logarithmic Model

The Logarithmic Model recognizes that if EBITDA is the product of revenue and EBITDA margin, they can be incorporated into the equity return multiplier expression for the Fund investment's Gross MOIC:

$$MOIC = \chi_{EG} \cdot \chi_{ME} \cdot \chi_{LEV} \cdot \chi_{FO}$$

$$= \frac{E_2}{E_1} \cdot \frac{M_2}{M_1} \cdot \frac{ER_2}{ER_1} \cdot \frac{FO_2}{FO_1} = \frac{(R_2 \cdot EM_2)}{(R_1 \cdot EM_1)} \cdot \frac{M_2}{M_1} \cdot \frac{ER_2}{ER_1} \cdot \frac{FO_2}{FO_1}$$

$$= \frac{R_2}{R_1} \cdot \frac{EM_2}{EM_1} \cdot \frac{M_2}{M_1} \cdot \frac{ER_2}{ER_1} \cdot \frac{FO_2}{FO_1}$$

$$\chi_{GROSS} = \chi_{RG} \cdot \chi_{EME} \cdot \chi_{ME} \cdot \chi_{LEV} \cdot \chi_{FO}$$

The Gross MOIC is equal to the product of the equity return multipliers for Revenue Growth (χ_{RG}), EBITDA Margin Expansion (χ_{EME}), Multiple Expansion (χ_{ME}), Leverage Effect (χ_{LEV}), and Fund Ownership Impact (χ_{FO}). The equity return multiples or multipliers for AlphaCo can be determined from the input variables above:

Table 8.3 AlphaCo value creation input variables

	t_1	t_2	X_2/X_1
TEV	\$100	\$180	1.80x
TEqV	\$80	\$180	2.25x
ND	\$20	\$0	n/a
ER	80%	100%	1.25x
E	\$20	\$30	1.50x
R	\$80	\$100	1.25x
EM	25%	30%	1.20x
M	5.0x	6.0x	1.20x
FEqV	\$44	\$88	2.00x
FO	55.0%	48.9%	0.89x

$$MOIC = \chi_{RG} \cdot \chi_{EME} \cdot \chi_{ME} \cdot \chi_{LEV} \cdot \chi_{FO}$$
$$= 1.25x \cdot 1.20x \cdot 1.20x \cdot 1.25x \cdot 0.89x = 2.00x$$

Because AlphaCo's R and EM values both increase, χ_{RG} and χ_{EME} must both be greater than 1.0x which increases the Gross MOIC and enhances value creation. They combine to produce the same equity return multiplier for EBITDA Growth, as shown below:

$$\chi_{EG} = \chi_{RG} \cdot \chi_{EME}$$
$$= 1.25x \cdot 1.20x = 1.50x$$

The other measures of value creation are described as follows. Value creation IRRs can be determined from equity return multipliers:

$$(1 + IRR_{GROSS}) = (1 + IRR_{RG}) \cdot (1 + IRR_{EME}) \cdot (1 + IRR_{ME}) \cdot (1 + IRR_{LEV}) \cdot (1 + IRR_{FO})$$

$$MOIC^{1/\Delta t} = \chi_{RG}{}^{1/\Delta t} \cdot \chi_{EME}{}^{1/\Delta t} \cdot \chi_{ME}{}^{1/\Delta t} \cdot \chi_{LEV}{}^{1/\Delta t} \cdot \chi_{FO}{}^{1/\Delta t}$$

Relative (%) value creation can be determined from equity return multipliers:

$$v_{RG} + v_{EME} + v_{ME} + v_{LEV} + v_{FO} = 100\%$$

$$\frac{\ln(\chi_{RG})}{\ln(MOIC)} + \frac{\ln(\chi_{EME})}{\ln(MOIC)} + \frac{\ln(\chi_{ME})}{\ln(MOIC)} + \frac{\ln(\chi_{LEV})}{\ln(MOIC)} + \frac{\ln(\chi_{FO})}{\ln(MOIC)} = 100\%$$

Solving for AlphaCo provides:

$$v_{RG} + v_{EME} + v_{ME} + v_{LEV} + v_{FO} = 100\%$$

$$\frac{\ln(1.25x)}{\ln(2.00x)} + \frac{\ln(1.20x)}{\ln(2.00x)} + \frac{\ln(1.20x)}{\ln(2.00x)} + \frac{\ln(1.25x)}{\ln(2.00x)} + \frac{\ln(0.89x)}{\ln(2.00x)} = 100\%$$

$$32.2\% + 26.3\% + 26.3\% + 32.2\% - 17.0\% = 100\%$$

Relative (%) value creation is additive, so the sum of v_{RG} and v_{EME} is the v_{EG} value of 58.5%. These relationships can be visualized with a stacked horizontal bar chart.

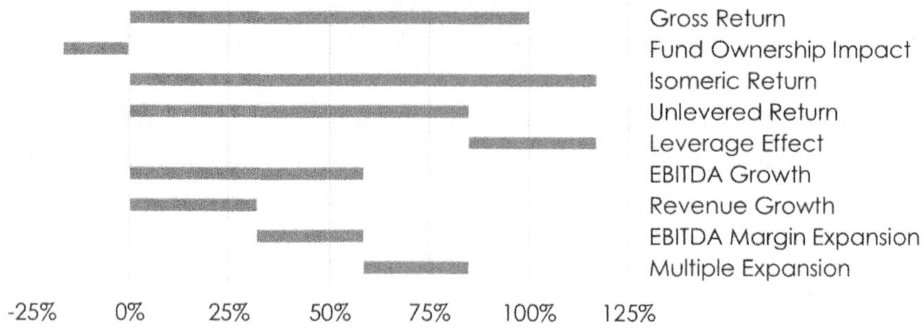

Figure 8.2 Logarithmic Model relative (%) value creation for AlphaCo

Absolute ($) value creation can be determined from relative (%) value creation:

$$\Delta FEqV = V_{RG} + V_{EME} + V_{ME} + V_{LEV} + V_{FO}$$
$$= v_{RG} \cdot \Delta FEqV + v_{EME} \cdot \Delta FEqV + v_{ME} \cdot \Delta FEqV + v_{LEV} \cdot \Delta FEqV + v_{FO} \cdot \Delta FEqV$$

Replacing EBITDA Growth with Revenue Growth and EBITDA Margin Expansion has no impact on the other value drivers, which are calculated as described in earlier chapters.

It is notable that the Revenue Growth formulas in this section are identical to those in Table 8.1. Unlike the Conventional and Derivative models, which require and additional factor ($\overline{EM}$) when measuring Revenue Growth for companies with EBITDA-based enterprise valuations, the Logarithmic Model formulas for Revenue Growth are independent of valuation methodology.

Further advantages of the Logarithmic Model become apparent as value creation is broken up into five or six components. The Derivative Model formulas were simpler in earlier chapters, but the number of input variables expand with each new value driver. The Derivative Model's Revenue Growth formula requires *FEqV*, *R*, *EM*, *M*, *ER*, and *FO*. In contrast, the Logarithmic Model's Revenue Growth formula only requires *FEqV* and *R*.

8.4. Model Comparisons

The following table compares the return measurements and value drivers provided by the Conventional, Derivative, and Logarithmic models. Results are presented in the form of absolute ($) value creation, relative (%) value creation, equity return multiples or multipliers, and value creation IRRs.

Table 8.4 Comparison of AlphaCo's value drivers

	Conv.	Deriv.	Log.		Conv.	Deriv.	Log.
V_{EG}	$28.6	$25.7	$25.7	x_{EG}	1.57x	1.50x	1.50x
V_{RG}	$15.7	$14.1	$14.2	x_{RG}	1.28x	1.25x	1.25x
V_{EME}	$12.9	$11.6	$11.6	x_{EME}	1.22x	1.20x	1.20x
V_{ME}	$13.0	$11.7	$11.6	x_{ME}	1.23x	1.20x	1.20x
$V_{CF/LEV}$	$10.4	$14.5	$14.2	$x_{CF/LEV}$	1.18x	1.26x	1.25x
V_{FO}	-$7.9	-$7.9	-$7.5	x_{FO}	0.88x	0.88x	0.89x
V_{GROSS}	$44.0	$44.0	$44.0	x_{GROSS}	2.00x	2.00x	2.00x
v_{EG}	64.9%	58.4%	58.5%	IRR_{EG}	11.9%	10.7%	10.7%
v_{RG}	35.7%	32.1%	32.2%	IRR_{RG}	6.4%	5.7%	5.7%
v_{EME}	29.2%	26.3%	26.3%	IRR_{EME}	5.2%	4.7%	4.7%
v_{ME}	29.5%	26.6%	26.3%	IRR_{ME}	5.2%	4.7%	4.7%
$v_{CF/LEV}$	23.6%	33.1%	32.2%	$IRR_{CF/LEV}$	4.2%	5.9%	5.7%
v_{FO}	-18.1%	-18.1%	-17.0%	IRR_{FO}	-3.1%	-3.1%	-2.9%
v_{GROSS}	100%	100%	100%	IRR_{GROSS}	18.9%	18.9%	18.9%

8.5. Summary

The three value creation models provide complete Fund-level descriptions of Revenue Growth, EBITDA Margin Expansion, Multiple Expansion, Leverage Effect or Cash Flow Generation, and Fund Ownership Impact. Each model can express returns and value drivers in the form of absolute ($) value creation, relative (%) value creation, equity return multiples or multipliers, and value creation IRRs.

As expected, Revenue Growth and EBITDA Margin Expansion values provided by the Derivative and Logarithmic models are nearly identical and slightly lower than values provided by the Conventional Model. The Conventional Model's tendency to understate the impact of leverage often leads to higher values for these components.

As value drivers are broken up into more numerous value creation components, the Logarithmic Model expressions maintain their basic structure, while the Derivative and Conventional model formulas become more complex and require additional value creation input variables. This trend continues in later chapters, as EBITDA Margin Expansion is broken into components of Gross Margin Expansion and Operating Margin Reduction, as Multiple Expansion is broken into components of Market Multiple Expansion and Intrinsic Multiple Expansion, and as Revenue Growth is broken into components of Addressable Market Growth and Market Share Growth.

The next chapter delves deeper into company P&Ls, specifically to examine the impact that company COGS and SG&A changes have on the value creation analysis.

9. Gross Profit Margins and Operating Expense Margins

This chapter describes EBITDA Margin Expansion in terms of its underlying components of Gross Margin Expansion and Operating Margin Reduction. Fund-level returns are expressed as the sum of Revenue Growth (V_{RG}), Gross Margin Expansion (V_{GME}), Operating Margin Reduction (V_{OMR}), Multiple Expansion (V_{ME}), Leverage Effect (V_{LEV}), and Fund Ownership Impact (V_{FO}), where:

$$\Delta FEqV = V_{RG} + V_{GME} + V_{OMR} + V_{ME} + V_{LEV} + V_{FO}$$

$$= \Delta R \cdot \overline{EM} \cdot \overline{M} \cdot \overline{ER} \cdot \overline{FO} + \Delta GM \cdot \overline{R} \cdot \overline{M} \cdot \overline{ER} \cdot \overline{FO} - \Delta OM \cdot \overline{R} \cdot \overline{M} \cdot \overline{ER} \cdot \overline{FO} + \Delta M \cdot \overline{E} \cdot \overline{ER} \cdot \overline{FO} + \Delta ER \cdot \overline{TEV} \cdot \overline{FO} + \Delta FO \cdot \overline{TEqV}$$

$$= \Delta FEqV \frac{\ln(R_2/R_1)}{\ln(MOIC)} + \Delta FEqV \frac{\Delta GM/\overline{EM}}{\ln(MOIC)} + \Delta FEqV \frac{-\Delta OM/\overline{EM}}{\ln(MOIC)} + \Delta FEqV \frac{\ln(M_2/M_1)}{\ln(MOIC)} + \Delta FEqV \frac{\ln(ER_2/ER_1)}{\ln(MOIC)} + \Delta FEqV \frac{\ln(FO_2/FO_1)}{\ln(MOIC)}$$

9.1. Background

EBITDA margin improvements are a good indication that a GP or management team is doing something right and having a positive impact on a private company investment. It is generally reasonable to give a GP full credit for EBITDA Margin Expansion when it happens.

Market forces do impact EBITDA margins. For example, changes can be driven by raw material costs that affect Cost of Goods Sold (COGS) or competition for talent which affects Selling, General, and Administrative Expenses (SG&A). While formulas for the market's impact on EBITDA margin, COGS, and SG&A can be derived, they are generally not practical or worth the effort.

Analyses of market-drivers require reasonable market proxies that reflect environmental conditions. Proxies must be transparent, uncontroversial, and span every potential holding period of interest. They must also be non-volatile. Any big jump from one quarter to the next generates noise and spurious results. Proxies should be easy to update and not require significant "massaging" every quarter to make them meaningful.

Chapter 10 and Chapter 11 describe the use of proxies for market valuation and addressable market size to evaluate their respective impact on Multiple Expansion and Revenue Growth. Market proxies work well in these areas, but meaningful measurements of "market" EBITDA margin, "market" COGS, or "market" SG&A are harder to justify and more difficult to maintain.

Furthermore, since the magnitudes of value creation components usually shrink as they become more refined, the downstream value drivers are generally less impactful. For example, in this chapter AlphaCo's gross profit margin improvements will generate a +1.8% value creation IRR. It is unlikely that a dissection of the market-driven and manager-driven forces behind +1.8% will provide additional meaningful insight into investment manager skill, especially if it is based on a dubious proxy for "market" COGS.

Just because an analysis can be done does not mean that it is worth the effort. In most cases, crediting the GP or management team with EBITDA Margin Expansion and its underlying components of Gross Margin Expansion and Operating Margin Reduction will provide a reasonable interpretation of results.

9.2. Conventional and Derivative Models

The rest of this chapter ignores the "market" and focuses entirely on company P&Ls. For private company investments, revenue (R) can be modeled as the sum of COGS, SG&A, and EBITDA (E). Subtracting COGS from R provides gross profit (GP), and subtracting both COGS and SG&A from R provides E:

$$E = R - COGS - SG\&A = GP - SG\&A$$

Dividing both sides by R reveals the company margins:

$$\frac{E}{R} = \frac{R - COGS}{R} - \frac{SG\&A}{R} = \frac{GP}{R} - \frac{SG\&A}{R}$$

$$EM = GM - OM$$

EBITDA margin (EM) is E/R, and gross profit margin (GM) is GP/R. The third term, $SG\&A/R$, is defined as the operating expense margin (OM). While it is not a common accounting term, OM makes intuitive sense and is necessary to separate EBITDA Margin Expansion into components. COGS and SG&A values are not usually explicitly provided in Fund reporting materials, but they are not needed if EM and GM can be determined from the P&L because the OM value is simply their difference.

Consider how value creation should relate to EM, GM, and OM:

- Higher EBITDA margins are better. Increasing EM means more E is generated for every dollar of R, which drives value creation.
- Higher gross profit margins are better. Increasing GM means COGS consumes less R, which increases E and drives value creation.
- Lower operating expense margins are better. Decreasing OM means SG&A consumes less R, which increases E and drives value creation.

If EBITDA margin is gross profit margin less operating expense margin, EBITDA margin changes can be expressed:

$$\Delta EM = EM_2 - EM_1$$
$$= (GM_2 - OM_2) - (GM_1 - OM_1) = (GM_2 - GM_1) - (OM_2 - OM_1)$$
$$= \Delta GM - \Delta OM$$

Consider a scenario where AlphaCo increases its gross profit margin from 60% to 62% as it increases its EBITDA margin from 25% to 30%. In this case, the operating expense margin decreases from OM_1 of 35% (60%-25%) to OM_2 of 32% (62%-30%), providing:

Table 9.1. AlphaCo value creation input variables

	t_1	t_2	ΔX	$\overline{X}$
TEV	\$100	\$180	+\$80	\$140
TEqV	\$80	\$180	+\$100	\$130
ND	\$20	\$0	-\$20	-
ER	80%	100%	+20%	90%
E	\$20	\$30	+\$10	\$25
M	5.0x	6.0x	+1.0x	5.5x
R	\$80	\$100	+\$20	\$90
EM	25%	30%	+5%	27.5%
GM	60%	62%	+2%	61%
OM	35%	32%	-3%	33.5%
FEqV	\$44	\$88	+44	-
FO	55.0%	48.9%	-6.1%	51.9%

Solving the formula for ΔEM provides:

$$\Delta EM = \Delta GM - \Delta OM$$
$$= 2.0\% - (-3.0\%) = 5.0\%$$

Since the expression $\Delta GM - \Delta OM$ is identical to $EM_2 - EM_1$, it can be substituted for ΔEM in the Conventional and Derivative model formulas for a Fund investment's EBITDA Margin Expansion (V_{EME}). The substitution creates two new value drivers: a term including ΔGM that represents Gross Margin Expansion (V_{GME}) and a term including $-\Delta OM$ that represents Operating Margin Reduction (V_{OMR}).

Conventional Model

Having established that Fund-level EBITDA Margin Expansion (V_{EME}) can be expressed:

$$V_{EME} = \Delta EM \cdot \overline{R} \cdot \overline{M} \cdot \overline{FO}$$

Replacing ΔEM with the expression above provides absolute (\$) value creation:

$$V_{EME} = (\Delta GM - \Delta OM) \cdot \overline{R} \cdot \overline{M} \cdot \overline{FO}$$
$$= \Delta GM \cdot \overline{R} \cdot \overline{M} \cdot \overline{FO} - \Delta OM \cdot \overline{R} \cdot \overline{M} \cdot \overline{FO}$$
$$= V_{GME} + V_{OMR}$$

Relative (%) value creation can be determined from absolute (\$) value creation:

$$v_{EME} = v_{GME} + v_{OMR}$$
$$\frac{V_{EME}}{\Delta FEqV} = \frac{V_{GME}}{\Delta FEqV} + \frac{V_{OMR}}{\Delta FEqV}$$

Equity return multipliers can be determined from relative (%) value creation:

$$x_{EME} = x_{GME} \cdot x_{OMR}$$
$$MOIC^{v_{EME}} = MOIC^{v_{GME}} \cdot MOIC^{v_{OMR}}$$

Value creation IRRs can be determined from equity return multipliers:

$$(1 + IRR_{EME}) = (1 + IRR_{GME}) \cdot (1 + IRR_{OMR})$$
$$x_{EME}{}^{1/\Delta t} = x_{GME}{}^{1/\Delta t} \cdot x_{OMR}{}^{1/\Delta t}$$

Replacing EBITDA Margin Expansion with Gross Margin Expansion and Operating Margin Reduction in the formulas above has no impact on the other value drivers, which are calculated as described in earlier chapters.

Derivative Model

Having established that Fund-level EBITDA Margin Expansion (V_{EME}) can be expressed:

$$V_{EME} = \Delta EM \cdot \overline{R} \cdot \overline{M} \cdot \overline{ER} \cdot \overline{FO}$$

Replacing ΔEM with the expression above provides absolute ($) value creation:

$$V_{EME} = (\Delta GM - \Delta OM) \cdot \overline{R} \cdot \overline{M} \cdot \overline{ER} \cdot \overline{FO}$$
$$= \Delta GM \cdot \overline{R} \cdot \overline{M} \cdot \overline{ER} \cdot \overline{FO} - \Delta OM \cdot \overline{R} \cdot \overline{M} \cdot \overline{ER} \cdot \overline{FO}$$
$$= V_{GME} + V_{OMR}$$

Solving for AlphaCo provides:

$$V_{EME} = V_{GME} + V_{OMR}$$
$$= 2\% \cdot \$90 \cdot 5.5\text{x} \cdot 90\% \cdot 51.9\% + 3\% \cdot \$90 \cdot 5.5\text{x} \cdot 90\% \cdot 51.9\%$$
$$= \$4.63 + \$6.94 = \$11.57$$

The value creation components V_{GME} and V_{OMR} can replace V_{EME} in the expression for $\Delta FEqV$ described in Chapter 8, providing:

$$\Delta FEqV = V_{RG} + V_{EME} + V_{ME} + V_{LEV} + V_{FO}$$
$$= V_{RG} + V_{GME} + V_{OMR} + V_{ME} + V_{LEV} + V_{FO}$$

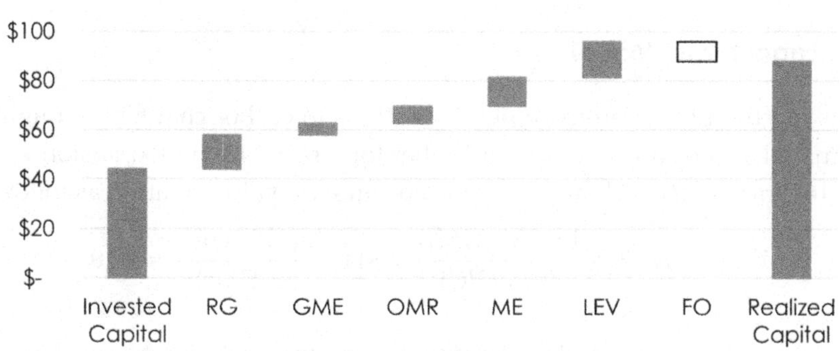

Figure 9.1 Derivative Model absolute ($) value creation bridge for AlphaCo

Relative (%) value creation can be determined from absolute ($) value creation:

$$v_{EME} = v_{GME} + v_{OMR}$$

$$\frac{V_{EME}}{\Delta FEqV} = \frac{V_{GME}}{\Delta FEqV} + \frac{V_{OMR}}{\Delta FEqV}$$

$$\frac{\$11.57}{\$44} = \frac{\$4.63}{\$44} + \frac{\$6.94}{\$44}$$

$$26.3\% = 10.5\% + 15.8\%$$

Equity return multipliers can be determined from relative (%) value creation:

$$x_{EME} = x_{GME} \cdot x_{OMR}$$

$$MOIC^{v_{EME}} = MOIC^{v_{GME}} \cdot MOIC^{v_{OMR}}$$

$$2.00x^{0.263} = 2.00x^{0.105} \cdot 2.00x^{0.158}$$

$$1.20x = 1.076x \cdot 1.116x$$

Value creation IRRs can be determined from equity return multipliers:

$$(1 + IRR_{EME}) = (1 + IRR_{GME}) \cdot (1 + IRR_{OMR})$$

$$x_{EME}{}^{1/\Delta t} = x_{GME}{}^{1/\Delta t} \cdot x_{OMR}{}^{1/\Delta t}$$

$$1.200x^{1/4.0} = 1.076x^{1/4.0} \cdot 1.116^{1/4.0}$$

$$1.047x = 1.018x \cdot 1.028x$$

$$(1 + 4.7\%) = (1 + 1.8\%) \cdot (1 + 2.8\%)$$

Replacing EBITDA Margin Expansion with Gross Margin Expansion and Operating Margin Reduction in the formulas above has no impact on the other value drivers, which are calculated as described in earlier chapters.

9.3. Logarithmic Model

Following the Logarithmic Model derivations in earlier chapters, it might be assumed that the equity return multiplier for Gross Margin Expansion would have the form GM_2/GM_1 and describe absolute ($) value creation as follows:

$$V_{GME} = \Delta FEqV \frac{\ln(GM_2/GM_1)}{\ln(MOIC)} = \$44 \cdot \frac{\ln(62\%/60\%)}{\ln(2.00x)} = \$2.08$$

This value is far from the $4.63 determined by the Derivative Model. Thus far, most Derivative and Logarithmic Model results have been within a few tenths of a percent. They are not in this case because x_{GME} does not equal GM_2/GM_1.

The Logarithmic Model excels at working with products, but *EM* is related to *GM* and *OM* by addition and subtraction. Recalling that:

$$\Delta EM = \Delta GM - \Delta OM$$
$$= 2.0\% - (-3.0\%) = 5.0\%$$

The 5.0% growth in EBITDA margin comes from a 2.0% increase in gross profit margin and a 3.0% decrease in operating expense margin. On a relative basis, the *GM* changes drive 40% of ΔEM (from $\Delta GM/\Delta EM$), and the *OM* changes drive 60% of ΔEM (from $-\Delta OM/\Delta EM$).

A rule of exponents states that X raised to the power of a sum, like $a + b$, must equal the product of X raised to the power of a and X raised to the power of b, so that:

$$X^{(a+b)} = X^a \cdot X^b$$

Since the relative drivers of ΔEM sum to 100%, it is reasonable to express the equity return multiplier relationship as follows:

$$\chi_{EME} = \chi_{GME} \cdot \chi_{OMR}$$
$${\chi_{EME}}^{\Delta EM/\Delta EM} = {\chi_{EME}}^{\Delta GM/\Delta EM} \cdot {\chi_{EME}}^{-\Delta OM/\Delta EM}$$
$$1.20x^{1.00} = 1.20x^{0.40} \cdot 1.20x^{0.60}$$
$$= 1.076x \cdot 1.116x = 1.200x$$

These equity return multipliers are almost identical to those provided by the Derivative Model and would provide nearly identical measurements of absolute ($) value creation, relative (%) value creation, and value creation IRR.

The only problem is that the formula is cumbersome. If χ_{GME} must be expressed as ${\chi_{EME}}^{\Delta GM/\Delta EM}$, the formula for V_{GME} becomes:

$$V_{GME} = \Delta FEqV \frac{\ln\left({\chi_{EME}}^{\Delta GM/\Delta EM}\right)}{\ln(MOIC)} = \$44 \cdot \frac{\ln(1.20x^{0.020/0.050})}{\ln(2.00x)} = \$4.63$$

The result is the same as the Derivative Model, but it requires working with a numerator that is a natural logarithm of a value whose exponent contains a fraction that includes differences. A simplification is suggested from the discussion of Derivative-Logarithmic equivalence in Section 5.8, where it is shown that the continuous change in X is:

$$r_X = \ln(X_2/X_1) \approx \frac{X_2 - X_1}{(X_1 + X_2)/2} = \frac{\Delta X}{\overline{X}}$$

The continuous change in EBITDA margin is, therefore:

$$\ln(EM_2/EM_1) \approx \frac{\Delta EM}{\overline{EM}} = \frac{\Delta GM - \Delta OM}{\overline{EM}} = \frac{\Delta GM}{\overline{EM}} + \frac{-\Delta OM}{\overline{EM}}$$

Substituting this expression into V_{EME} provides:

$$\begin{aligned} V_{EME} &= \Delta FEqV \frac{\ln(EM_2/EM_1)}{\ln(MOIC)} \approx \Delta FEqV \frac{\Delta EM/\overline{EM}}{\ln(MOIC)} \\ &= \Delta FEqV \frac{\Delta GM/\overline{EM}}{\ln(MOIC)} + \Delta FEqV \frac{-\Delta OM/\overline{EM}}{\ln(MOIC)} \\ &= V_{GME} + V_{OMR} \end{aligned}$$

Solving for AlphaCo provides:

$$\begin{gathered} V_{EME} = V_{GME} + V_{OMR} \\ = \Delta FEqV \frac{\Delta GM/\overline{EM}}{\ln(MOIC)} + \Delta FEqV \frac{-\Delta OM/\overline{EM}}{\ln(MOIC)} \\ \$44 \cdot \frac{5.0\%/27.5\%}{\ln(2.00x)} = \$44 \cdot \frac{2.0\%/27.5\%}{\ln(2.00x)} + \$44 \cdot \frac{3.0\%/27.5\%}{\ln(2.00x)} \\ \$11.54 = \$4.62 + \$6.92 \end{gathered}$$

The simplification provides values quite close to the Derivative Model and also points to solutions for χ_{GME} and χ_{OMR} that do not require first calculating χ_{EME}. The expression suggests that $\ln(\chi_{GME})$ is equivalent to $\Delta GM/\overline{EM}$ and $\ln(\chi_{OMR})$ is equivalent to $-\Delta OM/\overline{EM}$. Since e raised to the natural logarithm of X is equal to X:

$$\chi_{GME} = e^{\ln(\chi_{GME})} \approx e^{\Delta GM/\overline{EM}}$$

$$\chi_{OMR} = e^{\ln(\chi_{OMR})} \approx e^{-\Delta OM/\overline{EM}}$$

This formulation provides a direct estimate of the equity return multipliers for Gross Margin Expansion and Operating Margin Reduction:

$$\begin{gathered} \chi_{EME} = \chi_{GME} \cdot \chi_{OMR} \\ = e^{\Delta GM/\overline{EM}} \cdot e^{-\Delta OM/\overline{EM}} \\ = e^{0.020/0.275} \cdot e^{0.030/0.275} \\ = 1.075x \cdot 1.115x = 1.199x \end{gathered}$$

The results are not identical to earlier values, but variances will generally be tolerable in the domain of real world *EM*, *GM*, and *OM* changes. For example, consider an EM increase from 20% to 30%, which is quite substantial. The numerator of the V_{EME} and v_{EME} formulas would be:

$$\ln\left(\frac{EM_2}{EM_1}\right) = \ln\left(\frac{30\%}{20\%}\right) = 40.6\%$$

The simplified numerator of the V_{EME} and v_{EME} formulas would be quite close:

$$\ln\left(\frac{EM_2}{EM_1}\right) \approx \frac{\Delta EM}{\overline{EM}} = \frac{EM_2 - EM_1}{(EM_1 + EM_2)/2} = \frac{30\% - 20\%}{(20\% + 30\%)/2} = 40.0\%$$

With these definitions for χ_{GME} and χ_{OMR}, the Fund-level MOIC can be expressed:

$$\begin{aligned} MOIC &= \chi_{RG} \cdot \chi_{EME} \cdot \chi_{ME} \cdot \chi_{LEV} \cdot \chi_{FO} \\ &= \chi_{RG} \cdot \chi_{GME} \cdot \chi_{OMR} \cdot \chi_{ME} \cdot \chi_{LEV} \cdot \chi_{FO} \\ &= \frac{R_2}{R_1} \cdot \left(\frac{EM_2}{EM_1}\right)^{\Delta GM/\Delta EM} \cdot \left(\frac{EM_2}{EM_1}\right)^{-\Delta OM/\Delta EM} \cdot \frac{M_2}{M_1} \cdot \frac{ER_2}{ER_1} \cdot \frac{FO_2}{FO_1} \\ &\approx \frac{R_2}{R_1} \cdot e^{\Delta GM/\overline{EM}} \cdot e^{-\Delta OM/\overline{EM}} \cdot \frac{M_2}{M_1} \cdot \frac{ER_2}{ER_1} \cdot \frac{FO_2}{FO_1} \end{aligned}$$

The Gross MOIC is equal to the product of the equity return multipliers for Revenue Growth (χ_{RG}), Gross Margin Expansion (χ_{GME}), Operating Margin Reduction (χ_{OMR}), Multiple Expansion (χ_{ME}), Leverage Effect (χ_{LEV}), and Fund Ownership Impact (χ_{FO}).

AlphaCo's equity return multipliers are calculated from the input variables above and in Table 8.2:

$$\begin{aligned} MOIC &= \chi_{RG} \cdot \chi_{GME} \cdot \chi_{OMR} \cdot \chi_{ME} \cdot \chi_{LEV} \cdot \chi_{FO} \\ &= 1.250\text{x} \cdot 1.075\text{x} \cdot 1.115\text{x} \cdot 1.200\text{x} \cdot 1.250\text{x} \cdot 0.899\text{x} = 1.998\text{x} \end{aligned}$$

The other measures of value creation are described as follows. Value creation IRRs can be determined from equity return multipliers:

$$\begin{aligned} (1 + IRR_{GROSS}) = (1 + IRR_{RG}) \cdot (1 + IRR_{GME}) \cdot (1 + IRR_{OMR}) \cdot (1 + IRR_{ME}) \\ \cdot (1 + IRR_{LEV}) \cdot (1 + IRR_{FO}) \end{aligned}$$

$$MOIC^{1/\Delta t} = \chi_{RG}{}^{1/\Delta t} \cdot \chi_{GME}{}^{1/\Delta t} \cdot \chi_{OMR}{}^{1/\Delta t} \cdot \chi_{ME}{}^{1/\Delta t} \cdot \chi_{LEV}{}^{1/\Delta t} \cdot \chi_{FO}{}^{1/\Delta t}$$

Relative (%) value creation can be determined from equity return multipliers:

$$v_{RG} + v_{GME} + v_{OMR} + v_{ME} + v_{LEV} + v_{FO} = 100\%$$

$$\frac{\ln(\chi_{RG})}{\ln(MOIC)} + \frac{\Delta GM/\overline{EM}}{\ln(MOIC)} + \frac{-\Delta OM/\overline{EM}}{\ln(MOIC)} + \frac{\ln(\chi_{ME})}{\ln(MOIC)} + \frac{\ln(\chi_{LEV})}{\ln(MOIC)} + \frac{\ln(\chi_{FO})}{\ln(MOIC)} = 100\%$$

Solving for AlphaCo provides:

$$v_{RG} + v_{GME} + v_{OMR} + v_{ME} + v_{LEV} + v_{FO} = 100\%$$

$$\frac{\ln(1.25x)}{\ln(2.00x)} + \frac{2\%/27.5\%}{\ln(2.00x)} + \frac{3\%/27.5\%}{\ln(2.00x)} + \frac{\ln(1.20x)}{\ln(2.00x)} + \frac{\ln(1.25x)}{\ln(2.00x)} + \frac{\ln(0.89x)}{\ln(2.00x)} = 100\%$$

$$32.2\% + 10.5\% + 15.7\% + 26.3\% + 32.2\% - 17.0\% = 99.9\%$$

Relative (%) value creation is additive, so the sum of v_{GME} and v_{OMR} is the v_{EME} value of 26.3%. These relationships can be visualized with a stacked horizontal bar chart.

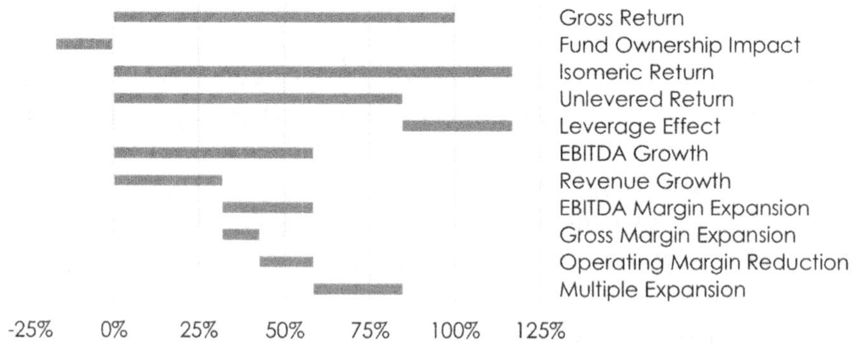

Figure 9.2 Logarithmic Model relative (%) value creation bridge for AlphaCo

Absolute ($) value creation can be determined from relative (%) value creation:

$$\Delta FEqV = V_{RG} + V_{GME} + V_{OMR} + V_{ME} + V_{LEV} + V_{FO}$$
$$= v_{RG} \cdot \Delta FEqV + v_{GME} \cdot \Delta FEqV + v_{OMR} \cdot \Delta FEqV + v_{ME} \cdot \Delta FEqV + v_{LEV} \cdot \Delta FEqV + v_{FO} \cdot \Delta FEqV$$

Replacing EBITDA Margin Expansion with Gross Margin Expansion and Operating Margin Reduction in the formulas above has no impact on the other value drivers, which are calculated as described in earlier chapters.

9.4. Model Comparisons

The following table compares the return measurements and value drivers provided by the Conventional, Derivative, and Logarithmic models. Results are presented in the form of absolute ($) value creation, relative (%) value creation, equity return multiples or multipliers, and value creation IRRs.

Table 9.2 Comparison of AlphaCo's value drivers

	Conv.	Deriv.	Log.		Conv.	Deriv.	Log.
V_{EG}	$28.6	$25.7	$25.7	X_{EG}	1.57x	1.50x	1.50x
V_{RG}	$15.7	$14.1	$14.2	X_{RG}	1.28x	1.25x	1.25x
V_{EME}	$12.9	$11.6	$11.6	X_{EME}	1.22x	1.20x	1.20x
V_{GME}	$5.1	$4.6	$4.6	X_{GME}	1.08x	1.08x	1.08x
V_{OMR}	$7.7	$6.9	$6.9	X_{OMR}	1.13x	1.12x	1.12x
V_{ME}	$13.0	$11.7	$11.6	X_{ME}	1.23x	1.20x	1.20x
$V_{CF/LEV}$	$10.4	$14.5	$14.2	$X_{CF/LEV}$	1.18x	1.26x	1.25x
V_{FO}	-$7.9	-$7.9	-$7.5	X_{FO}	0.88x	0.88x	0.89x
V_{GROSS}	$44.0	$44.0	$44.0	X_{GROSS}	2.00x	2.00x	2.00x
v_{EG}	64.9%	58.4%	58.5%	IRR_{EG}	11.9%	10.7%	10.7%
v_{RG}	35.7%	32.1%	32.2%	IRR_{RG}	6.4%	5.7%	5.7%
v_{EME}	29.2%	26.3%	26.3%	IRR_{EME}	5.2%	4.7%	4.7%
v_{GME}	11.7%	10.5%	10.5%	IRR_{GME}	2.0%	1.8%	1.8%
v_{OMR}	17.5%	15.8%	15.7%	IRR_{OMR}	3.1%	2.8%	2.8%
v_{ME}	29.5%	26.6%	26.3%	IRR_{ME}	5.2%	4.7%	4.7%
$v_{CF/LEV}$	23.6%	33.1%	32.2%	$IRR_{CF/LEV}$	4.2%	5.9%	5.7%
v_{FO}	-18.1%	-18.1%	-17.0%	IRR_{FO}	-3.1%	-3.1%	-2.9%
v_{GROSS}	100%	100%	100%	IRR_{GROSS}	18.9%	18.9%	18.9%

9.5. Summary

The three value creation models provide complete Fund-level descriptions of Revenue Growth, Gross Margin Expansion, Operating Margin Reduction, Multiple Expansion, Leverage Effect or Cash Flow Generation, and Fund Ownership Impact. Each model can express returns and value drivers in the form of absolute ($) value creation, relative (%) value creation, equity return multiples or multipliers, and value creation IRRs.

As expected, Gross Margin Expansion and Operating Margin Reduction values provided by the Derivative and Logarithmic models are nearly identical and slightly lower than values provided by the Conventional Model. The Conventional Model's tendency to understate the impact of leverage often leads to higher values for these components.

The next chapter begins to evaluate the impact of broader market forces on value creation analysis, starting with its influence on private company valuation multiples.

10. Market Multiples and Intrinsic Multiples

This chapter describes Multiple Expansion in terms of its underlying components of change in the market multiple (*MM*) and company valuation discount (or premium) to the market (μ). Fund-level returns are expressed as the sum of Revenue Growth (V_{RG}), EBITDA Margin Expansion (V_{EME}), Market Multiple Expansion (V_{MME}), Intrinsic Multiple Expansion (V_{IME}), Leverage Effect (V_{LEV}), and Fund Ownership Impact (V_{FO}), where:

$$\Delta FEqV = V_{RG} + V_{EME} + V_{MME} + V_{IME} + V_{LEV} + V_{FO}$$

$$= \Delta R \cdot \overline{EM} \cdot \overline{M} \cdot \overline{ER} \cdot \overline{FO} + \Delta EM \cdot \overline{R} \cdot \overline{M} \cdot \overline{ER} \cdot \overline{FO} + \Delta MM \cdot \overline{\mu} \cdot \overline{E} \cdot \overline{ER} \cdot \overline{FO} + \Delta \mu \cdot \overline{MM} \cdot \overline{E} \cdot \overline{ER} \cdot \overline{FO} + \Delta ER \cdot \overline{TEV} \cdot \overline{FO} + \Delta FO \cdot \overline{TEqV}$$

$$= \Delta FEqV \frac{\ln(R_2/R_1)}{\ln(MOIC)} + \Delta FEqV \frac{\ln(EM_2/EM_1)}{\ln(MOIC)} + \Delta FEqV \frac{\ln(MM_2/MM_1)}{\ln(MOIC)} + \Delta FEqV \frac{\ln(\mu_2/\mu_1)}{\ln(MOIC)} + \Delta FEqV \frac{\ln(ER_2/ER_1)}{\ln(MOIC)} + \Delta FEqV \frac{\ln(FO_2/FO_1)}{\ln(MOIC)}$$

10.1. Background

Some value creation models give GPs little or no credit for Multiple Expansion, which is considered to be a product of market timing or "luck". This oversimplification is unjust to many GPs that devote significant energies to sourcing, structuring, and negotiating investments with favorable valuations.

Broader market trends certainly impact valuation, but so does company performance. GPs should get credit for Multiple Expansion driven by companies becoming larger, achieving higher growth rates, or improving earnings quality. Furthermore, value can also be driven by buying a little lower or selling a little higher than "market", and approaches to sourcing, due diligence, managing acquisition or sale processes, and terms negotiations all impact price.

Private equity markets are presumed to be less efficient than public asset classes because buyers and sellers are fewer, transactions are more complex, and asset values are more opaque. Companies usually lack standardized reporting practices, significant finance teams, and transparent valuations. Private equity transactions must incur hefty due diligence costs to understand risks around company performance, customers, suppliers, competitors, regulators, etc. They also generate substantial legal costs as buyers and sellers spend months negotiating price and structuring terms. These factors contribute to the asset class's higher fees and higher return expectations.

Separating the market-driven and manager-driven components of Multiple Expansion requires first finding a reasonable proxy for market valuation. Proxies must be transparent, uncontroversial, and span every potential holding period of interest. They must be non-volatile because big jumps from one quarter to the next generate noise and spurious results. Proxies should also be easy to update and not require significant "massaging" every quarter to make them meaningful.

Fortunately, market valuation proxies are common in private equity. GPs and LPs talk about "market multiples" all the time. Investment bankers and service providers aggregate data for hundreds of private company M&A transactions each year and publish frequent reports with market multiples summarized by period, industry, geography, transaction size, etc. With a bit of effort, analysts should be able to compile enough data points to build reasonably representative sets of private company M&A multiples going back into the 1990s.

Market multiple data sets can also be built from public company valuations using market intelligence platforms like S&P CapitalIQ. Thanks to mutual fund and ETF managers, there are decades of available data for countless public equity strategies, and a reasonable index can be found to evaluate most Fund investment approaches.

10.2. Conventional and Derivative Models

Consider a scenario where prevailing market multiples (*MM*) increase from 7.0x to 8.0x, as AlphaCo's valuation multiple (*M*) increases from 5.0x to 6.0x. AlphaCo's discount (*d*) to the market is, therefore, 28.6% at t_1 and 25.0% at t_2.

The decrease in *d* suggests some measure of company-specific improvement. The company is worth 71.4% of *MM* at t_1 and then 75.0% of *MM* at t_2. Perhaps its growth rate increases, or it becomes more attractive to strategic buyers. *Private Equity Value Creation Analysis* measures valuation discount (or premium) to the market (μ) as $1 - d$, which is related to the valuation multiple and market multiple as follows:

$$\mu = 1 - d = M/MM$$

Table 10.1. AlphaCo value creation input variables

	t_1	t_2	ΔX	$\overline{X}$
TEV	\$100	\$180	+\$80	\$140
TEqV	\$80	\$180	+\$100	\$130
ND	\$20	\$0	-\$20	-
ER	80%	100%	+20%	90%
E	\$20	\$30	+\$10	\$25
M	5.0x	6.0x	+1.0x	5.5x
MM	7.0x	8.0x	+1.0x	7.5x
μ	71.4%	75.0%	+3.6%	73.2%
R	\$80	\$100	+\$20	\$90
EM	25%	30%	+5%	27.5%
FEqV	\$44	\$88	+44	-
FO	55.0%	48.9%	-6.1%	51.9%

Since the valuation multiple is the product of a market multiple and valuation discount, changes in *M* between times t_1 and t_2 are therefore:

$$\Delta M = M_2 - M_1$$
$$= MM_2 \cdot \mu_2 - MM_1 \cdot \mu_1$$

M, *MM*, and μ are all functions of time, so at any time *t*:

$$M_{(t)} = MM_{(t)} \cdot \mu_{(t)}$$

These terms can be expressed as time derivatives that are related by the "Product Rule":[7]

$$\frac{\partial M_{(t)}}{\partial t} = \frac{\partial}{\partial t}\left(MM_{(t)} \cdot \mu_{(t)}\right) = \frac{\partial MM_{(t)}}{\partial t} \cdot \mu_{(t)} + \frac{\partial \mu_{(t)}}{\partial t} \cdot MM_{(t)}$$

Replacing instantaneous changes (∂x) with discrete changes (Δx) and then multiplying both sides by Δt provides:

$$\Delta M = \Delta MM \cdot \overline{\mu} + \Delta \mu \cdot \overline{MM}$$

[7] A similar derivation, with additional steps and detail, is demonstrated in Section 4.2.

Solving ΔM for AlphaCo provides:

$$\begin{aligned}\Delta M &= \Delta MM \cdot \overline{\mu} + \Delta\mu \cdot \overline{MM}\\ &= 1.0\text{x} \cdot 73.2\% + 3.6\% \cdot 7.5\text{x}\\ &= 0.732\text{x} + 0.268\text{x} = 1.000\text{x}\end{aligned}$$

Since the expression $\Delta MM \cdot \overline{\mu} + \Delta\mu \cdot \overline{MM}$ is identical to $M_2 - M_1$, it can be substituted for ΔM in the Conventional and Derivative models of Fund-level Multiple Expansion (V_{ME}). The substitution creates two new value drivers: a term including $\Delta MM \cdot \overline{\mu}$ that represents Market Multiple Expansion (V_{MME}) and a term including $\Delta\mu \cdot \overline{MM}$ that represents Intrinsic Multiple Expansion (V_{IME}).

Conventional Model

Having established that Fund-level Multiple Expansion (V_{ME}) can be expressed:

$$V_{ME} = \Delta M \cdot \overline{E} \cdot \overline{FO}$$

Replacing ΔM with the expression above provides absolute ($) value creation:

$$\begin{aligned}V_{ME} &= (\Delta MM \cdot \overline{\mu} + \Delta\mu \cdot \overline{MM}) \cdot \overline{E} \cdot \overline{FO}\\ &= \Delta MM \cdot \overline{\mu} \cdot \overline{E} \cdot \overline{FO} + \Delta\mu \cdot \overline{MM} \cdot \overline{E} \cdot \overline{FO}\\ &= V_{MME} + V_{IME}\end{aligned}$$

Relative (%) value creation can be determined from absolute ($) value creation:

$$v_{ME} = v_{MME} + v_{IME}$$

$$\frac{V_{ME}}{\Delta FEqV} = \frac{V_{MME}}{\Delta FEqV} + \frac{V_{IME}}{\Delta FEqV}$$

Equity return multipliers can be determined from relative (%) value creation:

$$x_{ME} = x_{MME} \cdot x_{IME}$$

$$MOIC^{v_{ME}} = MOIC^{v_{MME}} \cdot MOIC^{v_{IME}}$$

Value creation IRRs can be determined from equity return multipliers:

$$(1 + IRR_{ME}) = (1 + IRR_{MME}) \cdot (1 + IRR_{IME})$$

$$x_{ME}^{\;1/\Delta t} = x_{MME}^{\;1/\Delta t} \cdot x_{IME}^{\;1/\Delta t}$$

Replacing Multiple Expansion with Market Multiple Expansion and Intrinsic Multiple Expansion in the formulas above has no impact on the other value drivers, which are calculated as described in earlier chapters.

Derivative Model

Having established that Fund-level Multiple Expansion can be expressed:

$$V_{ME} = \Delta M \cdot \overline{E} \cdot \overline{ER} \cdot \overline{FO}$$

Replacing ΔM with the expression above provides absolute ($) value creation:

$$V_{ME} = (\Delta MM \cdot \overline{\mu} + \Delta\mu \cdot \overline{MM}) \cdot \overline{E} \cdot \overline{ER} \cdot \overline{FO}$$
$$= \Delta MM \cdot \overline{\mu} \cdot \overline{E} \cdot \overline{ER} \cdot \overline{FO} + \Delta\mu \cdot \overline{MM} \cdot \overline{E} \cdot \overline{ER} \cdot \overline{FO}$$
$$= V_{MME} + V_{IME}$$

Solving for AlphaCo provides:

$$V_{ME} = V_{MME} + V_{IME}$$
$$= 1.0\text{x} \cdot 73.2\% \cdot \$25 \cdot 90\% \cdot 51.9\% + 3.6\% \cdot 7.5\text{x} \cdot \$25 \cdot 90\% \cdot 51.9\%$$
$$= \$8.56 + \$3.13 = \$11.69$$

The value creation components V_{MME} and V_{IME} can replace V_{ME} in the expression for $\Delta FEqV$ described in Chapter 8, providing:

$$\Delta FEqV = V_{RG} + V_{EME} + V_{ME} + V_{LEV} + V_{FO}$$
$$= V_{RG} + V_{EME} + V_{MME} + V_{IME} + V_{LEV} + V_{FO}$$

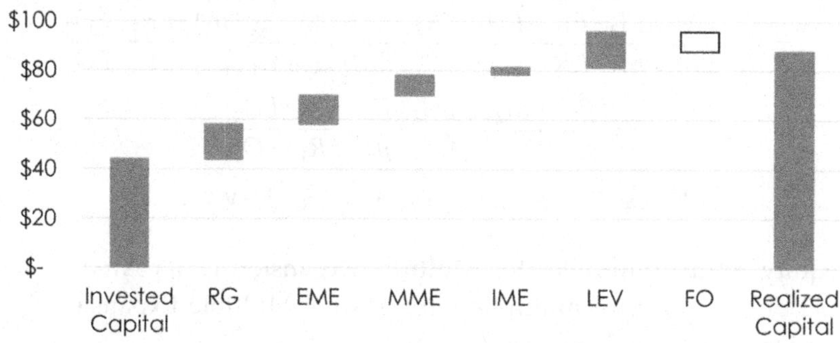

Figure 10.1 Derivative Model absolute ($) value creation bridge for AlphaCo

Relative (%) value creation can be determined from absolute ($) value creation:

$$v_{ME} = v_{MME} + v_{IME}$$

$$\frac{V_{ME}}{\Delta FEqV} = \frac{V_{MME}}{\Delta FEqV} + \frac{V_{IME}}{\Delta FEqV}$$

Equity return multipliers can be determined from relative (%) value creation:

$$\chi_{ME} = \chi_{MME} \cdot \chi_{IME}$$

$$MOIC^{v_{ME}} = MOIC^{v_{MME}} \cdot MOIC^{v_{IME}}$$

Value creation IRRs can be determined from equity return multipliers:

$$(1 + IRR_{ME}) = (1 + IRR_{MME}) \cdot (1 + IRR_{IME})$$

$$\chi_{ME}{}^{1/\Delta t} = \chi_{MME}{}^{1/\Delta t} \cdot \chi_{IME}{}^{1/\Delta t}$$

Replacing Multiple Expansion with Market Multiple Expansion and Intrinsic Multiple Expansion in the formulas above has no impact on the other value drivers, which are calculated as described in earlier chapters.

10.3. Logarithmic Model

The Logarithmic Model recognizes that if the valuation multiple is the product of a market multiple and a company valuation discount (or premium) to the market, they can be incorporated into the equity return multiplier expression for the Fund investment's Gross MOIC:

$$MOIC = \chi_{RG} \cdot \chi_{EME} \cdot \chi_{ME} \cdot \chi_{LEV} \cdot \chi_{FO}$$

$$= \frac{R_2}{R_1} \cdot \frac{EM_2}{EM_1} \cdot \frac{M_2}{M_1} \cdot \frac{ER_2}{ER_1} \cdot \frac{FO_2}{FO_1} = \frac{R_2}{R_1} \cdot \frac{EM_2}{EM_1} \cdot \frac{(MM_2 \cdot \mu_2)}{(MM_1 \cdot \mu_1)} \cdot \frac{ER_2}{ER_1} \cdot \frac{FO_2}{FO_1}$$

$$= \frac{R_2}{R_1} \cdot \frac{EM_2}{EM_1} \cdot \frac{MM_2}{MM_1} \cdot \frac{\mu_2}{\mu_1} \cdot \frac{ER_2}{ER_1} \cdot \frac{FO_2}{FO_1}$$

$$\chi_{GROSS} = \chi_{RG} \cdot \chi_{EME} \cdot \chi_{MME} \cdot \chi_{IME} \cdot \chi_{LEV} \cdot \chi_{FO}$$

The equity return multiplier for Multiple Expansion (χ_{ME}) is equal to the product of equity return multipliers for Market Multiple Expansion (χ_{MME}) and Intrinsic Multiple Expansion (χ_{IME}).

AlphaCo's equity return multipliers can be calculated from the input variables above:

Table 10.2 AlphaCo value creation input variables

	t_1	t_2	X_2/X_1
TEV	\$100	\$180	1.80x
$TEqV$	\$80	\$180	2.25x
ND	\$20	\$0	n/a
ER	80%	100%	1.25x
E	\$20	\$30	1.50x
R	\$80	\$100	1.25x
EM	25%	30%	1.20x
M	5.0x	6.0x	1.20x
MM	7.0x	8.0x	1.14x
μ	71.4%	75.0%	1.05x
$FEqV$	\$44	\$88	2.00x
FO	55.0%	48.9%	0.89x

Because AlphaCo's MM and μ values both increase, χ_{MME} and χ_{IME} must both be greater than 1.0x which increases the Fund's Gross MOIC and enhances value creation. They combine to produce the same equity multiplier for Multiple Expansion, as shown below:

$$\chi_{ME} = \chi_{MME} \cdot \chi_{IME}$$
$$= 1.143\text{x} \cdot 1.050\text{x} = 1.200\text{x}$$

The other measures of value creation are described as follows. Value creation IRRs can be determined from equity return multipliers:

$$(1 + IRR_{GROSS}) = (1 + IRR_{RG}) \cdot (1 + IRR_{EME}) \cdot (1 + IRR_{MME}) \cdot (1 + IRR_{IME}) \cdot (1 + IRR_{LEV}) \cdot (1 + IRR_{FO})$$

$$MOIC^{1/\Delta t} = \chi_{RG}{}^{1/\Delta t} \cdot \chi_{EME}{}^{1/\Delta t} \cdot \chi_{MME}{}^{1/\Delta t} \cdot \chi_{IME}{}^{1/\Delta t} \cdot \chi_{LEV}{}^{1/\Delta t} \cdot \chi_{FO}{}^{1/\Delta t}$$

Relative (%) value creation can be determined from equity return multipliers:

$$v_{RG} + v_{EME} + v_{MME} + v_{IME} + v_{LEV} + v_{FO} = 100\%$$

$$\frac{\ln(\chi_{RG})}{\ln(MOIC)} + \frac{\ln(\chi_{EME})}{\ln(MOIC)} + \frac{\ln(\chi_{MME})}{\ln(MOIC)} + \frac{\ln(\chi_{IME})}{\ln(MOIC)} + \frac{\ln(\chi_{LEV})}{\ln(MOIC)} + \frac{\ln(\chi_{FO})}{\ln(MOIC)} = 100\%$$

Solving for AlphaCo provides:

$$v_{RG} + v_{EME} + v_{MME} + v_{IME} + v_{LEV} + v_{FO} = 100\%$$

$$\frac{\ln(1.25x)}{\ln(2.00x)} + \frac{\ln(1.20x)}{\ln(2.00x)} + \frac{\ln(1.14x)}{\ln(2.00x)} + \frac{\ln(1.05x)}{\ln(2.00x)} + \frac{\ln(1.25x)}{\ln(2.00x)} + \frac{\ln(0.89x)}{\ln(2.00x)} = 100\%$$

$$32.2\% + 26.3\% + 19.3\% + 7.0\% + 32.2\% - 17.0\% = 100\%$$

Relative (%) value creation is additive, so the sum of v_{MME} and v_{IME} is the v_{ME} value of 26.3%. These relationships can be visualized with a stacked horizontal bar chart.

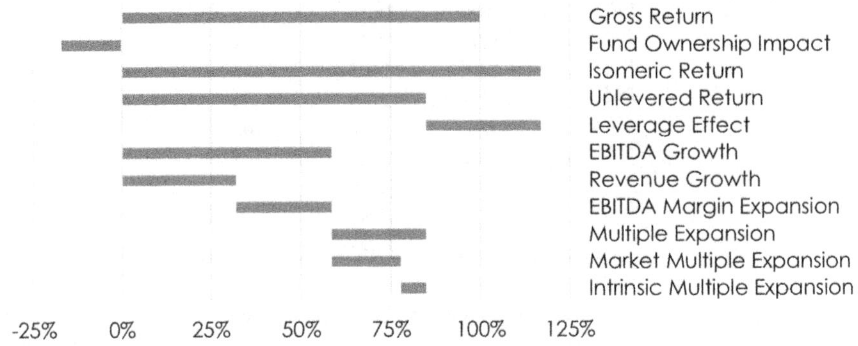

Figure 10.2 Logarithmic Model relative (%) value creation for AlphaCo

Absolute ($) value creation can be determined from relative (%) value creation:

$$\Delta FEqV = V_{RG} + V_{EME} + V_{MME} + V_{IME} + V_{LEV} + V_{FO}$$
$$= v_{RG} \cdot \Delta FEqV + v_{EME} \cdot \Delta FEqV + v_{MME} \cdot \Delta FEqV + v_{IME} \cdot \Delta FEqV + v_{LEV} \cdot \Delta FEqV + v_{FO} \cdot \Delta FEqV$$

Replacing Multiple Expansion with Market Multiple Expansion and Intrinsic Multiple Expansion in the formulas above has no impact on the other value drivers, which are calculated as described in earlier chapters.

10.4. Interpreting Intrinsic Multiple Expansion

Intrinsic Multiple Expansion must be interpreted differently for active and exited investments. One of the first principles of value creation analysis established in Chapter 3 is that equity values are assumed to be "real". It turns out that some equity values are more real than others.

At the moments of entry and exit, equity values are undeniably real. They match funds flowing between the bank accounts of buyers and sellers. Observers may opine that the buyer overpaid or underpaid for an asset, but wire transfers carry much more authority than the opinions of onlookers. As long as market valuation proxies are representative, Intrinsic Multiple Expansion should provide a meaningful description of reality.

In between entry and exit, it is possible for companies to be undervalued or overvalued. Real equity values exist, in theory, but are impossible to know precisely because of limited information about everything in the universe that may affect a company and the inability to process it. Equity values determined by GPs, therefore, are presumed to be imperfect estimates of an ideal value. Thus, for non-exited investments, Intrinsic Multiple Expansion is best used as an indicator of valuation appropriateness.

Consider a situation where an investment's valuation multiple increases by 20% while market multiples increase by 50%. This generates a χ_{ME} of 1.20x, χ_{MME} of 1.50x, and χ_{IME} of 0.80x. How should these numbers, which indicate negative Intrinsic Multiple Expansion, be interpreted? It could mean that the quality of company revenue or EBITDA has deteriorated materially since entry. Or it could simply mean that the GP is reluctant to write up the investment. GPs often take a conservative approach to active company valuations and hold them a bit lower than their anticipated exit values. Value creation analysis can quantify the magnitude of this discrepancy. If the valuation changes above occurred in AlphaCo, Intrinsic Multiple Expansion would be:

$$V_{IME} = \Delta FEqV \cdot \frac{\ln(\chi_{IME})}{\ln(MOIC)} = \$44 \cdot \frac{\ln(0.80x)}{\ln(2.00x)} = -\$14.2$$

Aggressive valuations can be quantified the same way. If the investment's valuation multiple increases by 50% while the market multiple increases by only 20%, then χ_{ME} is 1.50x, χ_{MME} is 1.20x, χ_{IME} is 1.25x, and Intrinsic Multiple Expansion is:

$$V_{IME} = \Delta FEqV \cdot \frac{\ln(\chi_{IME})}{\ln(MOIC)} = \$44 \cdot \frac{\ln(1.25x)}{\ln(2.00x)} = +\$14.2$$

The V_{IME} values above are useful but non-definitive. They are driven by two imperfect estimates of theoretically real values: an unrealized investment valuation and a market valuation proxy. Errors in either will accumulate in Intrinsic Multiple Expansion, so it tends to be more volatile than other value creation components. Nevertheless, it can be prudent to monitor V_{IME} changes because significant movements may indicate a need for further inquiry.

For exited investments, Intrinsic Multiple Expansion measurements are on firmer ground. The following circumstances can contribute to positive Intrinsic Multiple Expansion:

- Below market entry pricing, which can be driven by proprietary sourcing, clever structuring, purchase from a distressed seller, or investments in companies with certain "value inhibitors" that the GP can fix over its hold. These may include transitioning management, professionalizing businesses, addressing customer or sector concentration, regulatory issues, etc.

- Above market exit pricing, which can result from repositioning a company to provide a more valuable set of products or services, or by selling it to a strategic buyer that is willing to pay higher prices than other GPs.
- Market entry and exit pricing, but while achieving value creation through material business improvements, perhaps from a company becoming larger, achieving higher growth rates, or improving earnings quality.

For example, consider a directly sourced (non-intermediated) acquisition of a family-owned business that has customer concentration and needs a new management team. If the company goes on to make material business improvements, and then is sold to a strategic buyer for a premium value, all three circumstances would contribute to Intrinsic Multiple Expansion. Regardless of what happens to market multiples over the hold, Intrinsic Multiple Expansion would be quite strong, and a GP should get full credit for this result.

Circumstances that contribute to negative Intrinsic Multiple Expansion are basically the opposite, with negative value being driven by overpaying for businesses, managing them poorly, and running ineffective exit processes that lead to few potential buyers. GPs should get credit for this result as well.

For investments acquired from a GP, managed by a GP, and then sold to another GP, Intrinsic Multiple Expansion is usually limited to the third circumstance above. Pricing is closer to "market" when all buyers and sellers are non-distressed, professional investors. Investment bankers are involved at both entry and exit, which creates more "efficient" processes because many GPs have the opportunity to review and pursue the business.

By expressing value creation in terms of changes in the valuation discount (or premium) between entry and exit, the Intrinsic Multiple Expansion value creation component captures the combined impact of these various circumstances.

10.5. Model Comparisons

The following table compares the return measurements and value drivers provided by the Conventional, Derivative, and Logarithmic models. Results are presented in the form of absolute ($) value creation, relative (%) value creation, equity return multiples or multipliers, and value creation IRRs.

Table 10.3 Comparison of AlphaCo's value drivers

	Conv.	Deriv.	Log.		Conv.	Deriv.	Log.
V_{EG}	$28.6	$25.7	$25.7	X_{EG}	1.57x	1.50x	1.50x
V_{RG}	$15.7	$14.1	$14.2	X_{RG}	1.28x	1.25x	1.25x
V_{EME}	$12.9	$11.6	$11.6	X_{EME}	1.22x	1.20x	1.20x
V_{GME}	$5.1	$4.6	$4.6	X_{GME}	1.08x	1.08x	1.08x
V_{OMR}	$7.7	$6.9	$6.9	X_{OMR}	1.13x	1.12x	1.12x
V_{ME}	$13.0	$11.7	$11.6	X_{ME}	1.23x	1.20x	1.20x
V_{MME}	$9.5	$8.6	$8.5	X_{MME}	1.16x	1.14x	1.14x
V_{IME}	$3.5	$3.1	$3.1	X_{IME}	1.06x	1.05x	1.05x
$V_{CF/LEV}$	$10.4	$14.5	$14.2	$X_{CF/LEV}$	1.18x	1.26x	1.25x
V_{FO}	-$7.9	-$7.9	-$7.5	X_{FO}	0.88x	0.88x	0.89x
V_{GROSS}	$44	$44	$44	X_{GROSS}	2.00x	2.00x	2.00x
v_{EG}	64.9%	58.4%	58.5%	IRR_{EG}	11.9%	10.7%	10.7%
v_{RG}	35.7%	32.1%	32.2%	IRR_{RG}	6.4%	5.7%	5.7%
v_{EME}	29.2%	26.3%	26.3%	IRR_{EME}	5.2%	4.7%	4.7%
v_{GME}	11.7%	10.5%	10.5%	IRR_{GME}	2.0%	1.8%	1.8%
v_{OMR}	17.5%	15.8%	15.7%	IRR_{OMR}	3.1%	2.8%	2.8%
v_{ME}	29.5%	26.6%	26.3%	IRR_{ME}	5.2%	4.7%	4.7%
v_{MME}	21.6%	19.4%	19.3%	IRR_{MME}	3.8%	3.4%	3.4%
v_{IME}	7.9%	7.1%	7.0%	IRR_{IME}	1.4%	1.2%	1.2%
$v_{CF/LEV}$	23.6%	33.1%	32.2%	$IRR_{CF/LEV}$	4.2%	5.9%	5.7%
v_{FO}	-18.1%	-18.1%	-17.0%	IRR_{FO}	-3.1%	-3.1%	-2.9%
v_{GROSS}	100%	100%	100%	IRR_{GROSS}	18.9%	18.9%	18.9%

10.6. Summary

The three value creation models provide complete Fund-level descriptions of Revenue Growth, Gross Margin Expansion, Operating Margin Reduction, Market Multiple Expansion, Intrinsic Multiple Expansion, Leverage Effect or Cash Flow Generation, and Fund Ownership Impact. Each model can express returns and value drivers in the form of absolute ($) value creation, relative (%) value creation, equity return multiples or multipliers, and value creation IRRs.

As expected, Market Multiple Expansion and Intrinsic Multiple Expansion values provided by the Derivative and Logarithmic models are nearly identical and slightly lower than values provided by the Conventional Model. The Conventional Model's tendency to understate the impact of leverage often leads to higher values for these components.

The significance of any analysis that includes market-driven components is limited by the relevance of the underlying market proxies. Chapter 12 provides some examples of market valuation proxy data sets and describes various considerations for constructing them.

The next chapter examines the influence of another market force on value creation analysis, which is the impact of total addressable market changes on company revenue.

10.7. Optional: Other Market Components

This chapter's approach to separating the market-driven and manager-driven components of Multiple Expansion can also be applied to other value drivers like EBITDA Margin Expansion, Gross Margin Expansion, and Operating Margin Reduction.

The analyses require reasonable proxies for "market" EBITDA margin, "market" gross profit margin, and "market" operating expense margin. These are harder to build and maintain than market valuation proxies because even companies in the same industry can have quite different approaches to manufacturing or sales, which affect COGS or SG&A and lead to different ratios on the P&L. The data sets would require significant "massaging" every quarter to make them meaningful, and they would not remain applicable for very long before new entrants or M&A materially alter the market proxies.

However, assuming that reasonable market EBITDA margin (MEM) values could be determined, the company's relative discount (or premium) to the market (ε) would equal EM divided by MEM, providing:

$$\varepsilon = EM/MEM$$

Following earlier derivations, it can be shown that all variables are functions of time with relationships described by the Product Rule, which provides:

$$\Delta EM = \Delta MEM \cdot \overline{\varepsilon} + \Delta \varepsilon \cdot \overline{MEM}$$

Substituting this expression into the Conventional and Derivative model formulas for a Fund investment's EBITDA Margin Expansion (V_{EME}) creates two new value creation components: a term including $\Delta MEM \cdot \overline{\varepsilon}$ that represents Market EBITDA Margin Expansion (V_{MEME}) and a term including $\Delta \varepsilon \cdot \overline{MEM}$ that represents Intrinsic EBITDA Margin Expansion (V_{IEME}). In the Derivative Model, the formulas are:

$$V_{EME} = V_{MEME} + V_{IEME}$$
$$= \Delta MEM \cdot \overline{\varepsilon} \cdot \overline{R} \cdot \overline{M} \cdot \overline{ER} \cdot \overline{FO} + \Delta \varepsilon \cdot \overline{MEM} \cdot \overline{R} \cdot \overline{M} \cdot \overline{ER} \cdot \overline{FO}$$

Similarly, assuming that reasonable market gross profit margin (*MGM*) values could be determined, the company's relative discount (or premium) to the market (γ) would equal *GM* divided by *MGM*, providing:

$$\gamma = GM/MGM$$

Following earlier derivations, it can be shown that all variables are functions of time with relationships described by the Product Rule, which provides:

$$\Delta GM = \Delta MGM \cdot \overline{\gamma} + \Delta \gamma \cdot \overline{MGM}$$

Substituting this expression into the Conventional and Derivative model formulas for a Fund investment's Gross Margin Expansion (V_{GME}) creates two new value creation components: a term including $\Delta MGM \cdot \overline{\gamma}$ that represents Market Gross Margin Expansion (V_{MGME}) and a term including $\Delta \gamma \cdot \overline{MGM}$ that represents Intrinsic Gross Margin Expansion (V_{IGME}). In the Derivative Model, the formulas are:

$$V_{GME} = V_{MGME} + V_{IGME}$$
$$= \Delta MGM \cdot \overline{\gamma} \cdot \overline{R} \cdot \overline{M} \cdot \overline{ER} \cdot \overline{FO} + \Delta \gamma \cdot \overline{MGM} \cdot \overline{R} \cdot \overline{M} \cdot \overline{ER} \cdot \overline{FO}$$

Finally, if reasonable values for *MEM* and *MGM* are determined, a reasonable value for the market operating expense margin (*MOM*) may be defined by:

$$MOM = MGM - MEM$$

The company's relative discount (or premium) to the market (ω) would equal *OM* divided by *MOM*, providing:

$$\omega = OM/MOM$$

Following earlier derivations, it can be shown that all variables are functions of time with relationships described by the Product Rule, which provides:

$$\Delta OM = \Delta MOM \cdot \overline{\omega} + \Delta \omega \cdot \overline{MOM}$$

Substituting this expression into the Conventional and Derivative model formulas for a Fund investment's Operating Margin Reduction (V_{OMR}) creates two new value creation components: a term including $\Delta MOM \cdot \overline{\omega}$ that represents Market Operating Margin Reduction (V_{MOMR}) and a term including $\Delta \omega \cdot \overline{MOM}$ that represents Intrinsic Operating Margin Reduction (V_{IOMR}). In the Derivative Model, the formulas are:

$$V_{OMR} = V_{MOMR} + V_{IOMR}$$
$$= -\Delta MOM \cdot \overline{\omega} \cdot \overline{R} \cdot \overline{M} \cdot \overline{ER} \cdot \overline{FO} - \Delta \omega \cdot \overline{MOM} \cdot \overline{R} \cdot \overline{M} \cdot \overline{ER} \cdot \overline{FO}$$

Using the techniques described in previous chapters, derivations in this section can be adapted to also provide market-driven and manager-driven value creation component formulas in the Conventional and Logarithmic models of value creation.

The lack of widely applicable and representative market proxies for company margins limits the number of situations that can be practically evaluated with this approach. However, this technique may prove useful in specialized circumstances, like in the analysis of a specific company where industry-wide shifts in COGS and SG&A components are known to have had systemic P&L impacts on all company peers.

11. Addressable Market and Market Share

This chapter describes Revenue Growth in terms of changes in addressable market (β_R) and market share (α_R). Fund-level returns are expressed as the sum of Addressable Market Growth (V_{AMG}), Market Share Growth (V_{MSG}), EBITDA Margin Expansion (V_{EME}), Market Multiple Expansion (V_{MME}), Intrinsic Multiple Expansion (V_{IME}), Leverage Effect (V_{LEV}), and Fund Ownership Impact (V_{FO}), where:

$$\Delta FEqV = V_{AMG} + V_{MSG} + V_{EME} + V_{MME} + V_{IME} + V_{LEV} + V_{FO}$$

$$= \frac{\beta_R}{2}\left(R_1 + \frac{R_2}{1+\beta_R}\right)\overline{EM}\cdot\overline{M}\cdot\overline{ER}\cdot\overline{FO} + \frac{\alpha_R}{2}\left(R_1 + \frac{R_2}{1+\alpha_R}\right)\overline{EM}\cdot\overline{M}\cdot\overline{ER}\cdot\overline{FO}$$
$$+ \Delta EM\cdot\overline{R}\cdot\overline{M}\cdot\overline{ER}\cdot\overline{FO} + \Delta MM\cdot\overline{\mu}\cdot\overline{E}\cdot\overline{ER}\cdot\overline{FO}$$
$$+ \Delta\mu\cdot\overline{MM}\cdot\overline{E}\cdot\overline{ER}\cdot\overline{FO} + \Delta ER\cdot\overline{TEV}\cdot\overline{FO} + \Delta FO\cdot\overline{TEqV}$$

$$= \Delta FEqV\frac{\ln(1+\beta_R)}{\ln(MOIC)} + \Delta FEqV\frac{\ln(1+\alpha_R)}{\ln(MOIC)} + \Delta FEqV\frac{\ln(EM_2/EM_1)}{\ln(MOIC)}$$
$$+ \Delta FEqV\frac{\ln(MM_2/MM_1)}{\ln(MOIC)} + \Delta FEqV\frac{\ln(\mu_2/\mu_1)}{\ln(MOIC)}$$
$$+ \Delta FEqV\frac{\ln(ER_2/ER_1)}{\ln(MOIC)} + \Delta FEqV\frac{\ln(FO_2/FO_1)}{\ln(MOIC)}$$

11.1. Background

Like Multiple Expansion in the prior chapter, Revenue Growth can be broken into market-driven and manager-driven components. The latter is driven by successful sales execution, perhaps from acquiring new accounts, selling new products to existing accounts, improving customer retention, etc. While GPs use various KPIs to track sales function effectiveness, *Private Equity Value Creation Analysis* reduces the manager-driven component of Revenue Growth to simply growing sales faster than company peers.

The market-driven component of Revenue Growth, therefore, corresponds to the average sales growth of those peers. It accounts for the fact that companies selling goods or services into hot markets can grow revenue regardless of, or even despite of, directives from GPs and management teams. For expanding markets, the market-driven component would be positive, and a GP should only receive partial credit for company Revenue Growth. For contracting markets, the market-driven component would be negative, and a GP should earn extra credit for any company Revenue Growth achieved.

11.2. Conventional and Derivative Models

Similar to the derivation of Multiple Expansion components in Chapter 10, the company's relative discount (or premium) to the market would equal total company revenue (R) divided by a proxy for "market" revenue. This expression has a familiar form. If the proxy is defined as total addressable market (AM), the company-specific component would be market share (MS):

$$MS = R/AM$$

Since the revenue is the product of addressable market and market share, changes in R between times t_1 and t_2 are, therefore:

$$\begin{aligned}\Delta R &= R_2 - R_1 \\ &= MS_2 \cdot AM_2 - MS_1 \cdot AM_1\end{aligned}$$

R, AM, and MS are all functions of time, so at any time t:

$$R_{(t)} = AM_{(t)} \cdot MS_{(t)}$$

These terms can be expressed as time derivatives that are related by the "Product Rule":[8]

$$\frac{\partial R_{(t)}}{\partial t} = \frac{\partial}{\partial t}\left(AM_{(t)} \cdot MS_{(t)}\right) = \frac{\partial AM_{(t)}}{\partial t} \cdot MS_{(t)} + \frac{\partial MS_{(t)}}{\partial t} \cdot AM_{(t)}$$

Replacing instantaneous changes (∂x) with discrete changes (Δx) and then multiplying both sides by Δt provides:

$$\Delta R = \Delta AM \cdot \overline{MS} + \Delta MS \cdot \overline{AM}$$

For the value creation components described in prior chapters, the next step is to substitute this expression for ΔR into the formulas for a Fund

[8] A similar derivation, with additional steps and detail, is demonstrated in Section 4.2.

investment's Revenue Growth (V_{RG}). That is not recommended in this case because ΔAM and $\overline{AM}$ rely on the absolute ($) value of the total addressable market, which is a poor market proxy.

First, there is a problem of scale. If company revenue is $500 thousand and its addressable market is $5 billion, the AM value is $5,000,000,000 ($5 \times 10^9$) and the MS value is 0.01% (1×10^{-4}). Spreads of 10^{13} are common in chemistry and physics, but uncommon in finance. Rounding errors and mistakes caused by the misplacement of zeros and decimals are common when working with numbers that span such a large range. Furthermore, $5 billion is a suspiciously round number, and it is probably a poor representation of the market for AlphaCo's products and services. Addressable markets can be calculated a dozen ways with each providing different results. One analyst's $5 billion is another's $500 million, and values that differ by 10x fall short of the ideal market proxy, which should be transparent, uncontroversial, non-volatile, easy to update, etc.

Fortunately, value creation is driven by relative (%) change, not absolute ($) value. If the first analyst's addressable market grows from $5.0 billion to $5.5 billion, the other's would probably grow from $500 million to somewhere near $550 million. In both models, the addressable market should grow by around 10%, despite different underlying methods and assumptions. Reengineering the value creation formulas to depend on a relative (%) change in addressable market (β_R) makes them more robust, intuitive, and easier to work with. It is hard to visualize a billion of anything, but if a company's end-market grows by 5%, it makes sense that company revenue should also grow by at least 5% to keep up with its peers.

It also expands the universe of potentially useful value creation input variables. For example, an industry report might suggest that a particular market experienced a Compound Annual Growth Rate (CAGR) of 2.6% over some period. The report says nothing about the addressable market at times t_1 and t_2, but if they both fall within the period, and if the company is held for 4.0 years, then the market's β_R may be calculated as:

$$\beta_R = (1 + CAGR)^{\Delta t} - 1 = (1 + 2.6\%)^{4.0} - 1 = 10.8\%$$

GPs should have a general sense for the growth rates of their target markets. These are frequently reported by the media, industry groups, and analyst firms like Gartner and Forrester. They can also be estimated from groups of public company comparables. For example, the relative growth of the software market might be approximated by changes in total revenue for the S&P 500's 10 largest software companies. Relative market growth for companies providing products and services exclusively in Spain might be approximated by changes in the country's GDP.

Relative (%) change in addressable market (β_R) between times t_1 and t_2 can be expressed as follows:

$$\beta_R = \frac{AM_2 - AM_1}{AM_1} = \frac{AM_2}{AM_1} - 1$$

Similarly, the growth in market share (α_R), may be expressed:

$$\alpha_R = \frac{MS_2 - MS_1}{MS_1} = \frac{MS_2}{MS_1} - 1$$

It is useful to define changes in company P&Ls in the same format, where growth in company revenue (ρ_R) is:

$$\rho_R = \frac{R_2 - R_1}{R_1} = \frac{R_2}{R_1} - 1$$

Since MS at any time t can be expressed as R/AM, it can be substituted into the expression for α_R, providing:

$$\alpha_R = \frac{MS_2}{MS_1} - 1 = \frac{R_2/AM_2}{R_1/AM_1} - 1 = \frac{R_2/R_1}{AM_2/AM_1} - 1 = \frac{(1+\rho_R)}{(1+\beta_R)} - 1$$

This formula describes the relationships among α_R, β_R, and ρ_R. Since AlphaCo's revenue grows from \$80 to \$100, the formula provides a relative (%) change in revenue of 25%. If the relative (%) change in AlphaCo's addressable market is 10% over the period, then the relative (%) change in AlphaCo's market share is:

$$\alpha_R = \frac{(1+\rho_R)}{(1+\beta_R)} - 1 = \frac{(1+25\%)}{(1+10\%)} - 1 = 13.6\%$$

Returning to the expression for ΔR in the prior section, it is obvious that it includes one component driven by addressable market changes ($\Delta AM \cdot \overline{MS}$) and another component driven by market share changes ($\Delta MS \cdot \overline{AM}$):

$$\Delta R = \Delta AM \cdot \overline{MS} + \Delta MS \cdot \overline{AM}$$

The α_R and β_R variables can be incorporated into this expression by recognizing that AM_2 is $(1+\beta_R) \cdot AM_1$, and that MS_2 is $(1+\alpha_R) \cdot MS_1$.

Expanding $\Delta AM \cdot \overline{MS}$ and making the appropriate β_R substitutions provides:

$$\Delta AM \cdot \overline{MS} = (AM_2 - AM_1) \cdot \frac{1}{2}(MS_1 + MS_2)$$

$$= \frac{1}{2}(AM_2 \cdot MS_1 + AM_2 \cdot MS_2 - AM_1 \cdot MS_1 - AM_1 \cdot MS_2)$$

$$= \frac{1}{2}\left((1+\beta_R) \cdot AM_1 \cdot MS_1 + R_2 - R_1 - \frac{AM_2 \cdot MS_2}{(1+\beta_R)}\right)$$

$$= \frac{1}{2}\left(\beta_R \cdot R_1 + R_2\left(1 - \frac{1}{1+\beta_R}\right)\right)$$

$$= \frac{\beta_R}{2}\left(R_1 + \frac{R_2}{1+\beta_R}\right)$$

This expression is equivalent to the market-driven component of revenue change, $\Delta AM \cdot \overline{MS}$, and it only requires the input variables R_1, R_2, and β_R.

Similarly, expanding $\Delta MS \cdot \overline{AM}$ and making the appropriate α_R substitutions provides:

$$\Delta MS \cdot \overline{AM} = (MS_2 - MS_1) \cdot \frac{1}{2}(AM_1 + AM_2)$$

$$= \frac{1}{2}(MS_2 \cdot AM_1 + MS_2 \cdot AM_2 - MS_1 \cdot AM_1 - MS_1 \cdot AM_2)$$

$$= \frac{1}{2}\left((1+\alpha_R) \cdot MS_1 \cdot AM_1 + R_2 - R_1 - \frac{MS_2 \cdot AM_2}{(1+\alpha_R)}\right)$$

$$= \frac{1}{2}\left(\alpha_R \cdot R_1 + R_2\left(1 - \frac{1}{1+\alpha_R}\right)\right)$$

$$= \frac{\alpha_R}{2}\left(R_1 + \frac{R_2}{1+\alpha_R}\right)$$

This expression is equivalent to the manager-driven component of revenue change, $\Delta MS \cdot \overline{AM}$, and it only requires the input variables R_1, R_2, and α_R.

Solving for AlphaCo provides:

$$\Delta R = \Delta AM \cdot \overline{MS} + \Delta MS \cdot \overline{AM}$$

$$= \frac{\beta_R}{2}\left(R_1 + \frac{R_2}{1+\beta_R}\right) + \frac{\alpha_R}{2}\left(R_1 + \frac{R_2}{1+\alpha_R}\right)$$

$$= \frac{10\%}{2}\left(\$80 + \frac{\$100}{1+10\%}\right) + \frac{13.6\%}{2}\left(\$80 + \frac{\$100}{1+13.6\%}\right)$$

$$= \$8.55 + \$11.45 = \$20.00$$

Since the sum of growth rate expressions are identical to $R_2 - R_1$, they can be substituted for ΔR in the Conventional and Derivative model formulas for a

Fund investment's Revenue Growth (V_{RG}). The substitution creates two new value drivers: a term including β_R that represents Addressable Market Growth (V_{AMG}) and a term including α_R that represents Market Share Growth (V_{MSG}).

Conventional Model

Having established that Fund-level Revenue Growth (V_{RG}) can be expressed:

$$V_{RG} = \Delta R \cdot \overline{EM} \cdot \overline{M} \cdot \overline{FO}$$

Replacing ΔR with the expression above provides absolute (\$) value creation:

$$V_{RG} = \frac{\beta_R}{2}\left(R_1 + \frac{R_2}{1+\beta_R}\right)\overline{EM} \cdot \overline{M} \cdot \overline{FO} + \frac{\alpha_R}{2}\left(R_1 + \frac{R_2}{1+\alpha_R}\right)\overline{EM} \cdot \overline{M} \cdot \overline{FO}$$
$$= V_{AMG} + V_{MSG}$$

Relative (%) value creation can be determined from absolute (\$) value creation:

$$v_{RG} = v_{AMG} + v_{MSG}$$
$$\frac{V_{RG}}{\Delta FEqV} = \frac{V_{AMG}}{\Delta FEqV} + \frac{V_{MSG}}{\Delta FEqV}$$

Equity return multipliers can be determined from relative (%) value creation:

$$\chi_{RG} = \chi_{AMG} \cdot \chi_{MSG}$$
$$MOIC^{v_{RG}} = MOIC^{v_{AMG}} \cdot MOIC^{v_{MSG}}$$

Value creation IRRs can be determined from equity return multipliers:

$$(1 + IRR_{RG}) = (1 + IRR_{AMG}) \cdot (1 + IRR_{MSG})$$
$${\chi_{RG}}^{1/\Delta t} = {\chi_{AMG}}^{1/\Delta t} \cdot {\chi_{MSG}}^{1/\Delta t}$$

Replacing Revenue Growth with Addressable Market Growth and Market Share Growth in the formulas above has no impact on the other value drivers, which are calculated as described in earlier chapters.

Derivative Model

Having established that Fund-level Revenue Growth (V_{RG}) can be expressed:

$$V_{RG} = \Delta R \cdot \overline{EM} \cdot \overline{M} \cdot \overline{ER} \cdot \overline{FO}$$

Replacing ΔR with the expression above provides absolute ($) value creation:

$$V_{RG} = \frac{\beta_R}{2}\left(R_1 + \frac{R_2}{1+\beta_R}\right)\overline{EM}\cdot\overline{M}\cdot\overline{ER}\cdot\overline{FO} + \frac{\alpha_R}{2}\left(R_1 + \frac{R_2}{1+\alpha_R}\right)\overline{EM}\cdot\overline{M}\cdot\overline{ER}\cdot\overline{FO}$$
$$= V_{AMG} + V_{MSG}$$

Solving for AlphaCo provides:

$$V_{RG} = V_{AMG} + V_{MSG}$$
$$= \frac{10\%}{2}\left(\$80 + \frac{\$100}{1+10\%}\right)\cdot 27.5\%\cdot 5.5\text{x}\cdot 90\%\cdot 51.9\%$$
$$+ \frac{13.6\%}{2}\left(\$80 + \frac{\$100}{1+13.6\%}\right)\cdot 27.5\%\cdot 5.5\text{x}\cdot 90\%\cdot 51.9\%$$
$$= \$6.04 + \$8.10 = \$14.14$$

The value creation components V_{AMG} and V_{MSG} can replace V_{RG} in the expression for $\Delta FEqV$ described in Chapter 10, providing:

$$\Delta FEqV = V_{RG} + V_{EME} + V_{MME} + V_{IME} + V_{LEV} + V_{FO}$$
$$= V_{AMG} + V_{MSG} + V_{EME} + V_{MME} + V_{IME} + V_{LEV} + V_{FO}$$

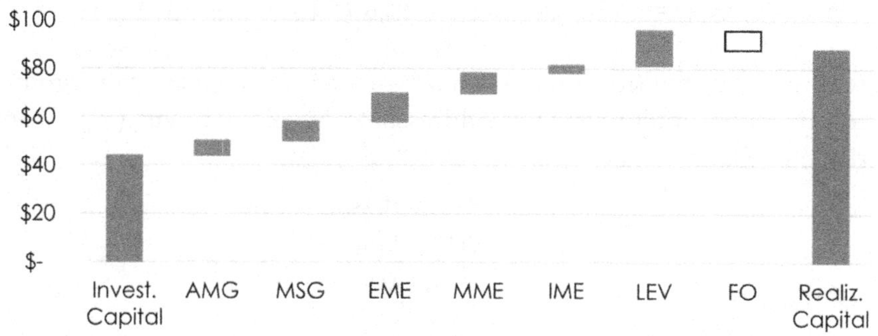

Figure 11.1 Derivative Model absolute ($) value creation bridge for AlphaCo

Relative (%) value creation can be determined from absolute ($) value creation:

$$v_{RG} = v_{AMG} + v_{MSG}$$
$$\frac{V_{RG}}{\Delta FEqV} = \frac{V_{AMG}}{\Delta FEqV} + \frac{V_{MSG}}{\Delta FEqV}$$

Equity return multipliers can be determined from relative (%) value creation:

$$\chi_{RG} = \chi_{AMG}\cdot\chi_{MSG}$$
$$MOIC^{v_{RG}} = MOIC^{v_{AMG}}\cdot MOIC^{v_{MSG}}$$

Value creation IRRs can be determined from equity return multipliers:

$$(1 + IRR_{RG}) = (1 + IRR_{AMG}) \cdot (1 + IRR_{MSG})$$

$$\chi_{RG}{}^{1/\Delta t} = \chi_{AMG}{}^{1/\Delta t} \cdot \chi_{MSG}{}^{1/\Delta t}$$

Replacing Revenue Growth with Addressable Market Growth and Market Share Growth in the formulas above has no impact on the other value drivers, which are calculated as described in earlier chapters.

11.3. Logarithmic Model

The Logarithmic Model recognizes that if revenue is the product of addressable market and market share, they can be incorporated into the equity return multiplier expression for the Fund investment's Gross MOIC:

$$MOIC = \chi_{RG} \cdot \chi_{EME} \cdot \chi_{MME} \cdot \chi_{IME} \cdot \chi_{LEV} \cdot \chi_{FO}$$

$$= \frac{R_2}{R_1} \cdot \frac{EM_2}{EM_1} \cdot \frac{MM_2}{MM_1} \cdot \frac{\mu_2}{\mu_1} \cdot \frac{ER_2}{ER_1} \cdot \frac{FO_2}{FO_1} = \frac{(AM_2 \cdot MS_2)}{(AM_1 \cdot MS_1)} \cdot \frac{EM_2}{EM_1} \cdot \frac{MM_2}{MM_1} \cdot \frac{\mu_2}{\mu_1} \cdot \frac{ER_2}{ER_1} \cdot \frac{FO_2}{FO_1}$$

$$= \frac{AM_2}{AM_1} \cdot \frac{MS_2}{MS_1} \cdot \frac{EM_2}{EM_1} \cdot \frac{MM_2}{MM_1} \cdot \frac{\mu_2}{\mu_1} \cdot \frac{ER_2}{ER_1} \cdot \frac{FO_2}{FO_1}$$

$$\chi_{GROSS} = \chi_{AMG} \cdot \chi_{MSG} \cdot \chi_{EME} \cdot \chi_{MME} \cdot \chi_{IME} \cdot \chi_{LEV} \cdot \chi_{FO}$$

The equity return multiplier for Revenue Growth (χ_{RG}) is equal to the product of equity return multipliers for Addressable Market Growth (χ_{AMG}) and Market Share Growth (χ_{MSG}).

$$\chi_{RG} = \chi_{AMG} \cdot \chi_{MSG}$$

$$\frac{R_2}{R_1} = \frac{AM_2}{AM_1} \cdot \frac{MS_2}{MS_1}$$

Note how the equity return multipliers relate to the relative (%) change in revenue, addressable market, and market share:

$$\rho_R = \frac{R_2}{R_1} - 1 = \chi_{RG} - 1$$

$$\beta_R = \frac{AM_2}{AM_1} - 1 = \chi_{AMG} - 1$$

$$\alpha_R = \frac{MS_2}{MS_1} - 1 = \chi_{MSG} - 1$$

Substituting these values into the formula for equity return multipliers provides:

$$\chi_{RG} = \chi_{AMG} \cdot \chi_{MSG}$$
$$(1 + \rho_R) = (1 + \beta_R) \cdot (1 + \alpha_R)$$
$$(1 + 25.0\%) = (1 + 10.0\%) \cdot (1 + 13.6\%)$$
$$1.25x = 1.10x \cdot 1.14x$$

These relative (%) change expressions provide results that are identical to the formulas with equity return multipliers. Substituting the growth rate expressions for AM_2/AM_1 and MS_2/MS_1 provides:

$$MOIC = \chi_{AMG} \cdot \chi_{MSG} \cdot \chi_{EME} \cdot \chi_{MME} \cdot \chi_{IME} \cdot \chi_{LEV} \cdot \chi_{FO}$$
$$= (1 + \beta_R) \cdot (1 + \alpha_R) \cdot \frac{EM_2}{EM_1} \cdot \frac{MM_2}{MM_1} \cdot \frac{\mu_2}{\mu_1} \cdot \frac{ER_2}{ER_1} \cdot \frac{FO_2}{FO_1}$$

The other measures of value creation are described as follows. Value creation IRRs can be determined from equity return multipliers:

$$(1 + IRR_{GROSS}) = (1 + IRR_{AMG}) \cdot (1 + IRR_{MSG}) \cdot (1 + IRR_{EME}) \cdot (1 + IRR_{MME}) \cdot (1 + IRR_{IME}) \cdot (1 + IRR_{LEV}) \cdot (1 + IRR_{FO})$$

$$MOIC^{1/\Delta t} = \chi_{AMG}{}^{1/\Delta t} \cdot \chi_{MSG}{}^{1/\Delta t} \cdot \chi_{EME}{}^{1/\Delta t} \cdot \chi_{MME}{}^{1/\Delta t} \cdot \chi_{IME}{}^{1/\Delta t} \cdot \chi_{LEV}{}^{1/\Delta t} \cdot \chi_{FO}{}^{1/\Delta t}$$

Relative (%) value creation can be determined from equity return multipliers:

$$v_{AMG} + v_{MSG} + v_{EME} + v_{MME} + v_{IME} + v_{LEV} + v_{FO} = 100\%$$
$$\frac{\ln(\chi_{AMG})}{\ln(MOIC)} + \frac{\ln(\chi_{MSG})}{\ln(MOIC)} + \frac{\ln(\chi_{EME})}{\ln(MOIC)} + \frac{\ln(\chi_{MME})}{\ln(MOIC)} + \frac{\ln(\chi_{IME})}{\ln(MOIC)} + \frac{\ln(\chi_{LEV})}{\ln(MOIC)} + \frac{\ln(\chi_{FO})}{\ln(MOIC)} = 100\%$$

Solving for AlphaCo provides:

$$v_{AMG} + v_{MSG} + v_{EME} + v_{MME} + v_{IME} + v_{LEV} + v_{FO} = 100\%$$
$$\frac{\ln(1 + 10.0\%)}{\ln(2.00x)} + \frac{\ln(1 + 13.6\%)}{\ln(2.00x)} + \frac{\ln(1.20x)}{\ln(2.00x)} + \frac{\ln(1.14x)}{\ln(2.00x)} + \frac{\ln(1.05x)}{\ln(2.00x)} + \frac{\ln(1.25x)}{\ln(2.00x)} + \frac{\ln(0.89x)}{\ln(2.00x)} = 100\%$$
$$13.8\% + 18.4\% + 26.3\% + 19.3\% + 7.0\% + 32.2\% - 17.0\% = 100\%$$

Relative (%) value creation is additive, so the sum of v_{AMG} and v_{MSG} is the v_{RG} value of 32.2%. These relationships can be visualized with a stacked horizontal bar chart.

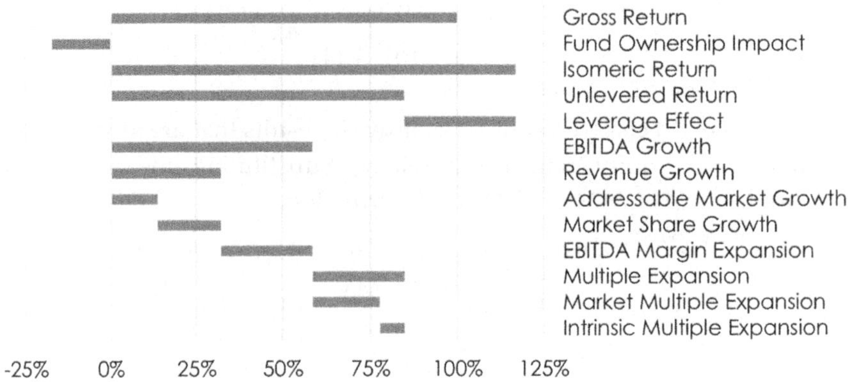

Figure 11.2 Logarithmic Model relative (%) value creation for AlphaCo

Absolute ($) value creation can be determined from relative (%) value creation:

$$\Delta FEqV = V_{AMG} + V_{MSG} + V_{EME} + V_{MME} + V_{IME} + V_{LEV} + V_{FO}$$
$$= v_{AMG} \cdot \Delta FEqV + v_{MSG} \cdot \Delta FEqV + v_{EME} \cdot \Delta FEqV + v_{MME} \cdot \Delta FEqV + v_{IME} \cdot \Delta FEqV + v_{LEV} \cdot \Delta FEqV + v_{FO} \cdot \Delta FEqV$$

Replacing Revenue Growth with Addressable Market Growth and Market Share Growth in the formulas above has no impact on the other value drivers, which are calculated as described in earlier chapters.

11.4. Model Comparisons

The following table compares the return measurements and value drivers provided by the Conventional, Derivative, and Logarithmic models. Results are presented in the form of absolute ($) value creation, relative (%) value creation, equity return multiples or multipliers, and value creation IRRs.

Table 11.1 Comparison of AlphaCo's value drivers

	Conv.	Deriv.	Log.		Conv.	Deriv.	Log.
V_{EG}	$28.6	$25.7	$25.7	X_{EG}	1.57x	1.50x	1.50x
V_{RG}	$15.7	$14.1	$14.2	X_{RG}	1.28x	1.25x	1.25x
V_{AMG}	$6.7	$6.0	$6.1	X_{AMG}	1.11x	1.10x	1.10x
V_{MSG}	$9.0	$8.1	$8.1	X_{MSG}	1.15x	1.14x	1.14x
V_{EME}	$12.9	$11.6	$11.6	X_{EME}	1.22x	1.20x	1.20x
V_{GME}	$5.1	$4.6	$4.6	X_{GME}	1.08x	1.08x	1.08x
V_{OMR}	$7.7	$6.9	$6.9	X_{OMR}	1.13x	1.12x	1.12x
V_{ME}	$13.0	$11.7	$11.6	X_{ME}	1.23x	1.20x	1.20x
V_{MME}	$9.5	$8.6	$8.5	X_{MME}	1.16x	1.14x	1.14x
V_{IME}	$3.5	$3.1	$3.1	X_{IME}	1.06x	1.05x	1.05x
$V_{CF/LEV}$	$10.4	$14.5	$14.2	$X_{CF/LEV}$	1.18x	1.26x	1.25x
V_{FO}	-$7.9	-$7.9	-$7.5	X_{FO}	0.88x	0.88x	0.89x
V_{GROSS}	$44	$44	$44	X_{GROSS}	2.00x	2.00x	2.00x
v_{EG}	64.9%	58.4%	58.5%	IRR_{EG}	11.9%	10.7%	10.7%
v_{RG}	35.7%	32.1%	32.2%	IRR_{RG}	6.4%	5.7%	5.7%
v_{AMG}	15.3%	13.7%	13.8%	IRR_{AMG}	2.7%	2.4%	2.4%
v_{MSG}	20.5%	18.4%	18.4%	IRR_{MSG}	3.6%	3.2%	3.2%
v_{EME}	29.2%	26.3%	26.3%	IRR_{EME}	5.2%	4.7%	4.7%
v_{GME}	11.7%	10.5%	10.5%	IRR_{GME}	2.0%	1.8%	1.8%
v_{OMR}	17.5%	15.8%	15.7%	IRR_{OMR}	3.1%	2.8%	2.8%
v_{ME}	29.5%	26.6%	26.3%	IRR_{ME}	5.2%	4.7%	4.7%
v_{MME}	21.6%	19.4%	19.3%	IRR_{MME}	3.8%	3.4%	3.4%
v_{IME}	7.9%	7.1%	7.0%	IRR_{IME}	1.4%	1.2%	1.2%
$v_{CF/LEV}$	23.6%	33.1%	32.2%	$IRR_{CF/LEV}$	4.2%	5.9%	5.7%
v_{FO}	-18.1%	-18.1%	-17.0%	IRR_{FO}	-3.1%	-3.1%	-2.9%
v_{GROSS}	100%	100%	100%	IRR_{GROSS}	18.9%	18.9%	18.9%

11.5. Summary

The three value creation models provide complete Fund-level descriptions of Addressable Market Growth, Market Share Growth, Gross Margin Expansion, Operating Margin Reduction, Market Multiple Expansion, Intrinsic Multiple Expansion, Leverage Effect or Cash Flow Generation, and Fund Ownership Impact. Each model can express returns and value drivers in the form of absolute ($) value creation, relative (%) value creation, equity return multiples or multipliers, and value creation IRRs.

As expected, Addressable Market Growth and Market Share Growth values provided by the Derivative and Logarithmic models are nearly identical and slightly lower than values provided by the Conventional Model. The Conventional Model's tendency to understate the impact of leverage often leads to higher values for these components.

The significance of any analysis that includes market-driven components is limited by the relevance of the underlying market proxies. Chapter 12 provides some examples of addressable market proxy data sets and describes various considerations for constructing them. It will also combine Market Multiple Expansion and Addressable Market Growth to describe a Market-driven Return that functions much like PME.

12. Market-driven and Manager-driven Returns

This chapter describes Fund-level value creation in terms of its underlying components of Market-driven Return (V_{MARKET}) and Manager-driven Return ($V_{MANAGER}$), where:

$$\Delta FEqV = V_{MARKET} + V_{MANAGER}$$

$$= \left\{ \left[\frac{\beta_R}{2} \left(R_1 + \frac{R_2}{1+\beta_R} \right) \overline{EM} \cdot \overline{M} + \Delta MM \cdot \overline{\mu} \cdot \overline{E} \right] \cdot \overline{ER} \cdot \overline{FO} \right\}$$

$$+ \left\{ \left[\frac{\alpha_R}{2} \left(R_1 + \frac{R_2}{1+\alpha_R} \right) \overline{EM} \cdot \overline{M} + \Delta EM \cdot \overline{R} \cdot \overline{M} + \Delta\mu \cdot \overline{MM} \cdot \overline{E} \right] \cdot \overline{ER} \cdot \overline{FO} + \Delta ER \cdot \overline{TEV} \cdot \overline{FO} + \Delta FO \cdot \overline{TEqV} \right\}$$

$$= \Delta FEqV \frac{\ln\left((1+\beta_R) \cdot MM_2/MM_1\right)}{\ln(MOIC)} + \Delta FEqV \left[1 - \frac{\ln\left((1+\beta_R) \cdot MM_2/MM_1\right)}{\ln(MOIC)} \right]$$

$$= \Delta FEqV \frac{\sum r_{MARKET}}{\ln(MOIC)} + \Delta FEqV \left(1 - \frac{\sum r_{MARKET}}{\ln(MOIC)} \right)$$

12.1. Background

A measurement of market returns termed Public Market Equivalence, or PME, is briefly described in Chapter 1. The technique replaces private company or portfolio cash flows with investments in an index like the S&P 500 or Russell 3000. PME represents the Fund or private company investment's opportunity cost, which is a return that could have been achieved with cheaper and more liquid investment alternatives. Market Multiple Expansion and Addressable Market Growth, described in prior chapters, perform a similar task, and when they are combined they can provide a measure that is much like PME.

12.2. Market-driven Return

Combining Market Multiple Expansion (V_{MME}) and Addressable Market Growth (V_{AMG}) provides a Market-driven Return (V_{MARKET}), where:

$$V_{MARKET} = V_{AMG} + V_{MME}$$

The Conventional Model would define market-driven absolute ($) value creation as:

$$\begin{aligned} V_{MARKET} &= \frac{\beta_R}{2}\left(R_1 + \frac{R_2}{1+\beta_R}\right)\overline{EM}\cdot\overline{M}\cdot\overline{FO} + \Delta MM\cdot\overline{\mu}\cdot\overline{E}\cdot\overline{FO} \\ &= \left[\frac{\beta_R}{2}\left(R_1 + \frac{R_2}{1+\beta_R}\right)\overline{EM}\cdot\overline{M} + \Delta MM\cdot\overline{\mu}\cdot\overline{E}\right]\cdot\overline{FO} \end{aligned}$$

The Derivative Model would define market-driven absolute ($) value creation as:

$$\begin{aligned} V_{MARKET} &= \frac{\beta_R}{2}\left(R_1 + \frac{R_2}{1+\beta_R}\right)\overline{EM}\cdot\overline{M}\cdot\overline{ER}\cdot\overline{FO} + \Delta MM\cdot\overline{\mu}\cdot\overline{E}\cdot\overline{ER}\cdot\overline{FO} \\ &= \left[\frac{\beta_R}{2}\left(R_1 + \frac{R_2}{1+\beta_R}\right)\overline{EM}\cdot\overline{M} + \Delta MM\cdot\overline{\mu}\cdot\overline{E}\right]\cdot\overline{ER}\cdot\overline{FO} \end{aligned}$$

Consistent with the value creation components described in prior chapters, the Conventional and Derivative models differ by a factor equal to the average equity ratio ($\overline{ER}$).

Solving for AlphaCo using the Derivative Model provides:

$$V_{MARKET} = V_{AMG} + V_{MME}$$

$$= \left[\frac{10\%}{2}\left(\$80 + \frac{\$100}{1+10\%}\right)27.5\%\cdot 5.5\text{x} + 1.0\text{x}\cdot 73.2\%\cdot \$25\right]\cdot 90\%\cdot 51.9\%$$

$$= \$14.6$$

It is a perfectly accurate expression, but the Derivative Model not only requires details about market growth and market multiples, it also needs company-specific data for revenue, EBITDA, EBITDA margin, enterprise valuation, market valuation discount, debt levels, and Fund ownership percentages.

In contrast, the Logarithmic Model requires only details about the return and market proxies. It recognizes that the Market-driven Return multiple (χ_{MARKET}) must be the product of the equity return multipliers for Addressable Market Growth (χ_{AMG}) and Market Multiple Expansion (χ_{MME}), where:

$$\chi_{MARKET} = \chi_{AMG} \cdot \chi_{MME} = (1 + \beta_R) \cdot \frac{MM_2}{MM_1}$$

Solving for AlphaCo provides:

$$\chi_{MARKET} = (1 + 10\%) \cdot \frac{8.0x}{7.0x} = 1.26x$$

This asserts that the market environment – in the form of 10% addressable market growth and expansion of the market multiple from 7.0x to 8.0x – would have driven a 1.26x Gross MOIC for any investment in the space. A Fund investment that returns 1.26x creates no intrinsic value, while those that return more or less than 1.26x either outperform or underperform the "market".

It is quite a useful expression because it is independent of any company-specific ownership, performance, or valuation details. It does not matter if the Fund owns 5% or 95% of a company, or if it is a cash-burning growth recapitalization or a control buyout that is levered to the hilt. If reasonably representative proxies for the addressable market and market multiples can be determined, the market proxies and basic investment return numbers like Gross MOIC are all that is needed to produce a PME alternative.

The other measures of value creation are described as follows. Market-driven IRRs can be determined from equity return multipliers:

$$IRR_{MARKET} = \chi_{MARKET}{}^{1/\Delta t} - 1$$
$$= \left((1 + \beta_R) \cdot MM_2/MM_1\right)^{1/\Delta t} - 1$$
$$= 1.26x^{1/4.0} - 1 = 5.8\%$$

Market-driven relative (%) value creation can be determined from equity return multipliers:

$$\upsilon_{MARKET} = \frac{\ln(\chi_{MARKET})}{\ln(MOIC)}$$
$$= \frac{\ln\left((1 + \beta_R) \cdot MM_2/MM_1\right)}{\ln(MOIC)}$$
$$= \frac{\ln(1.26x)}{\ln(2.00x)} = 33.0\%$$

Market-driven absolute ($) value creation can be determined from relative (%) value creation:

$$V_{MARKET} = \Delta FEqV \cdot v_{MARKET}$$

$$= \Delta FEqV \cdot \frac{\ln\left((1+\beta_R)\cdot MM_2/MM_1\right)}{\ln(MOIC)}$$

$$= \$44 \cdot 33.0\% = \$14.5$$

Since the Logarithmic and Derivative models produce nearly identical results, the Logarithmic Model is generally preferred. It does not require company-specific value creation input variables that may be difficult to consistently obtain or verify. Basic return measurements like Gross MOIC, which are almost always available in Fund reporting materials, are the only company-specific input variables needed to complete the analysis.

12.3. Manager-driven Return

Defining V_{AMG} and V_{MME} as components of V_{MARKET} implies that the rest of value creation can be attributed to the GP or investment manager ($V_{MANAGER}$). The Derivative and Logarithmic models would both define $V_{MANAGER}$ as the sum of Market Share Growth (V_{MSG}), EBITDA Margin Expansion (V_{EME}), Intrinsic Multiple Expansion (V_{IME}), Leverage Effect (V_{LEV}), and Fund Ownership Impact (V_{FO}):

$$V_{MANAGER} = V_{MSG} + V_{EME} + V_{IME} + V_{LEV} + V_{FO}$$

$$= \left[\frac{\alpha_R}{2}\left(R_1 + \frac{R_2}{1+\alpha_R}\right)\overline{EM}\cdot\overline{M} + \Delta EM\cdot\overline{R}\cdot\overline{M} + \Delta\mu\cdot\overline{MM}\cdot\overline{E}\right]\cdot\overline{ER}\cdot\overline{FO}$$
$$+ \Delta ER\cdot\overline{TEV}\cdot\overline{FO} + \Delta FO\cdot\overline{TEqV}$$

Similarly, the Manager-driven return multiple ($\chi_{MANAGER}$) can be defined as the product of the equity return multipliers for these components, where:

$$\chi_{MANAGER} = \chi_{MSG}\cdot\chi_{EME}\cdot\chi_{IME}\cdot\chi_{LEV}\cdot\chi_{FO}$$

Solving for AlphaCo provides:

$$\chi_{MANAGER} = (1+\alpha_R)\cdot\frac{EM_2}{EM_1}\cdot\frac{\mu_2}{\mu_1}\cdot\frac{ER_2}{ER_1}\cdot\frac{FO_2}{FO_1}$$

$$= (1+13.6\%)\cdot\frac{30\%}{25\%}\cdot\frac{75.0\%}{71.4\%}\cdot\frac{100\%}{80\%}\cdot\frac{48.9\%}{55.0\%}$$

$$= 1.14x\cdot 1.20x\cdot 1.05x\cdot 1.25x\cdot 0.89x = 1.59x$$

The Fund's investment in AlphaCo outperformed the "market" because its Manager-driven Return multiple is greater than 1.00x. This is expected because the Fund investment's Gross MOIC is 2.00x, which is higher than the Market-driven Return multiple of 1.26x.

The same result can be achieved without company-specific details. Because Gross MOIC is the product of the Market-driven Return multiple and the Manager-driven Return multiple, $\chi_{MANAGER}$ can be expressed:

$$\chi_{MANAGER} = \frac{MOIC}{\chi_{MARKET}} = \frac{MOIC}{(1+\beta_R)\cdot MM_2/MM_1}$$
$$= \frac{2.00x}{1.26x} = 1.59x$$

The other measures of value creation are described as follows. Manager-driven IRRs can be determined from equity return multipliers:

$$IRR_{MANAGER} = \chi_{MANAGER}{}^{1/\Delta t} - 1$$
$$= \left(\frac{MOIC}{(1+\beta_R)\cdot MM_2/MM_1}\right)^{1/\Delta t} - 1$$
$$= 1.59x^{1/4.0} - 1 = 12.3\%$$

Manager-driven relative (%) value creation can be determined from equity return multipliers:

$$v_{MANAGER} = \frac{\ln(\chi_{MANAGER})}{\ln(MOIC)}$$
$$= \frac{\ln\left(\frac{MOIC}{(1+\beta_R)\cdot MM_2/MM_1}\right)}{\ln(MOIC)} = \frac{\ln(MOIC) - \ln\left((1+\beta_R)\cdot MM_2/MM_1\right)}{\ln(MOIC)}$$
$$= 100\% - \frac{\ln\left((1+\beta_R)\cdot MM_2/MM_1\right)}{\ln(MOIC)}$$
$$= 100\% - 33.0\% = 67.0\%$$

Manager-driven absolute ($) value creation can be determined from relative (%) value creation:

$$V_{MANAGER} = \Delta FEqV \cdot v_{MANAGER}$$
$$= \Delta FEqV \cdot \left[100\% - \frac{\ln\left((1+\beta_R)\cdot MM_2/MM_1\right)}{\ln(MOIC)}\right]$$
$$= \$44 \cdot 67.0\% = \$29.5$$

The Logarithmic Model provides a robust Fund-level description of market-adjusted returns. Unlike the Derivative Model, it requires no company-specific ownership, performance, or valuation details. Unlike PME, it requires no index time series data or detailed company cash flows. Model results are only limited by the ability to find reasonable proxies for market valuations and addressable markets. Examples of such market proxies are provided in Section 12.5 and Section 12.6.

12.4. Model Comparisons

The following table compares the Market-driven Return and Manager-driven Return provided by the Conventional, Derivative, and Logarithmic models. Results are presented in the form of absolute ($) value creation, relative (%) value creation, equity return multiples or multipliers, and value creation IRRs.

Table 12.1 Comparison of AlphaCo's value drivers

	Conv.	Deriv.	Log.		Conv.	Deriv.	Log.
V_{MARKET}	*$16.2*	*$14.6*	*$14.5*	X_{MARKET}	*1.29x*	*1.26x*	*1.26x*
$V_{MANAGER}$	*$27.8*	*$29.4*	*$29.5*	$X_{MANAGER}$	*1.55x*	*1.59x*	*1.59x*
V_{GROSS}	*$44.0*	*$44.0*	*$44.0*	X_{GROSS}	*2.00x*	*2.00x*	*2.00x*
v_{MARKET}	*36.9%*	*33.2%*	*33.0%*	IRR_{MARKET}	*6.6%*	*5.9%*	*5.9%*
$v_{MANAGER}$	*63.1%*	*66.8%*	*67.0%*	$IRR_{MANAGER}$	*11.6%*	*12.3%*	*12.3%*
v_{GROSS}	*100%*	*100%*	*100%*	IRR_{GROSS}	*18.9%*	*18.9%*	*18.9%*

12.5. Example: US Middle Market Manufacturing

Consider the market proxies for a US Middle-Market manufacturing company. Research firms and investment banks track private company M&A transactions and occasionally distribute valuation metrics through publications like Pitchbook's *US PE Middle Market Report* and Baird's *Annual M&A Outlook*. Such reports can provide multiples by geography, industry, or transaction size. The figures in Table 12.2 are from Pitchbook's 2019 annual US PE Middle Market Report (Pitchbook 2020).

Table 12.2 Private company transaction EBITDA multiples

	2010	2011	2012	2013	2014	2015	2016	2017	2018	2019
All Private Equity (PE) Buyouts	8.0x	8.8x	8.4x	10.1x	12.8x	10.0x	10.9x	11.9x	11.5x	12.5x
Middle Market PE Buyouts	8.1x	9.2x	8.7x	10.2x	11.4x	10.0x	8.5x	11.6x	11.5x	12.8x

The EBITDA multiples above are not manufacturing-specific, but they correspond to the right geography (US) and transaction size (Middle Market). Consider a Middle Market investment made in 2013 and exited in 2017. EBITDA multiples increase from 10.2x to 11.6x over the four-year hold. The table suggests that value creation driven by Market Multiple Expansion can be represented by an equity return multiplier of 1.13x and value creation IRR of 3.2%.

Industry-specific research reports or economic data might be used to provide market growth estimates. The US Bureau of Economic Analysis tracks various measures of economic activity including Gross Output by industry. Gross Output values are not assumed to be the same as the addressable market, but if Gross Output for an industry changes by 10%, it might be assumed that β_R for a company operating in that industry might also change by about 10%. Table 12.3 provides Gross Output for total US manufacturing and the 11 industries that make up total US durable goods manufacturing (US Bureau of Economic Analysis 2020).

Table 12.3 US manufacturing industry Gross Output ($billions)

	2010	2011	2012	2013	2014	2015	2016	2017	2018	2019
Total US manufacturing	5,019	5,593	5,771	5,939	6,026	5,706	5,551	5,819	6,217	6,266
Total US durable goods mfg.	2,383	2,599	2,700	2,790	2,906	2,902	2,847	2,969	3,168	3,242
Wood products	72	73	81	91	98	101	106	113	116	114
Nonmetallic mineral products	92	94	100	106	114	119	125	129	133	138
Primary metals	236	282	269	263	266	227	205	224	256	246
Fabricated metal products	296	327	344	351	361	352	337	364	396	407
Machinery	319	369	405	397	409	379	351	381	400	409
Computer & electronic products	359	361	354	353	348	348	341	353	377	396
Elec. equip., appliances, comps.	109	118	123	124	126	126	124	129	136	140
Vehicles, bodies, trailers, parts	435	498	513	565	629	677	701	711	754	784
Other transportation equipment	249	259	291	306	325	333	316	320	345	348
Furniture & related products	57	57	66	67	69	74	74	76	78	81
Miscellaneous manufacturing	160	162	156	168	163	166	167	171	178	180

Gross Output for Machinery Manufacturing is $396.8 billion in 2013 and $380.5 billion in 2017. If the change is a reasonable proxy for market growth, β_R would be -4.1%. This suggests that value creation driven by Addressable Market Growth can be represented by an equity return multiplier of 0.96x and value creation IRR of -1.0%. Since the market proxy decreases over the hold, any company that grows revenue, or even loses less than 4.1% of their revenue, would beat the "market" and create value through Market Share Growth.

Market Multiple Expansion and Addressable Market Growth combine to produce a Market-driven Return multiple that is the product of 1.13x and 0.96x, or 1.09x, which suggests a Market-driven IRR of 2.0%. If a Fund investment in the space generates a 2.00x Gross MOIC and 18.9% Gross IRR, the market-adjusted, Manager-driven Return multiple would be 2.00x/1.09x, or 1.84x, and the Manager-driven IRR would be 16.5%. The investment manager would be responsible for 88% of value creation.

A Market Return Matrix can provide a useful snapshot of return expectations across various entry and exit date combinations. The Pitchbook Middle-Market PE multiples in Table 12.2 and US Machinery Manufacturing Gross Output values in Table 12.3 combine to provide the following:

Table 12.4 US Middle Market manufacturing Return Matrix by entry and exit year

	Exit or Current Year									
Entry Year	2010	2011	2012	2013	2014	2015	2016	2017	2018	2019
2010	1.00x	1.31x	1.36x	1.57x	1.81x	1.46x	1.15x	1.70x	1.78x	2.02x
2011	-	1.00x	1.04x	1.20x	1.38x	1.12x	0.87x	1.30x	1.36x	1.54x
2012	-	-	1.00x	1.15x	1.32x	1.07x	0.84x	1.25x	1.30x	1.48x
2013	-	-	-	1.00x	1.15x	0.93x	0.73x	1.09x	1.13x	1.29x
2014	-	-	-	-	1.00x	0.81x	0.63x	0.94x	0.98x	1.12x
2015	-	-	-	-	-	1.00x	0.78x	1.17x	1.22x	1.38x
2016	-	-	-	-	-	-	1.00x	1.49x	1.55x	1.76x
2017	-	-	-	-	-	-	-	1.00x	1.04x	1.19x
2018	-	-	-	-	-	-	-	-	1.00x	1.14x
2019	-	-	-	-	-	-	-	-	-	1.00x

Table 12.4 suggests that representative US Middle Market Machine Manufacturing deals should generate Gross MOICs of at least 1.81x if held from 2010 to 2014, 0.84x if held from 2012 to 2016, 0.98x if held from 2014 to 2018, and so on. The matrix provides a useful representation of Market-driven Returns that can fit onto an index card.

Of course, the technique is only effective if the market proxies are accurate and appropriate to the companies being analyzed. In this case, the market valuation multiples may be suspect. Valuation details for private company transactions are "private". Research firms and investment banks do an admirable job collecting as much data as possible, but it is a small percentage of the universe. The Pitchbook Middle Market PE multiples in Table 12.2 are based on 269 datapoints, about 27 deals per year. In the same report, Pitchbook estimates the total number of Middle Market PE transactions for the period to be over 21 thousand, or about 2,100 deals per year.

The next section looks at market proxies built from public company data. This is a trade-off. Private company transaction data is more relevant to private company transactions, but data quality is poorer. Public company data is less relevant to private company transactions, but data quality is much higher.

12.6. Example: Software Market

Public company data sets are larger and more reliable, but they must be kept as simple as possible. Consider a software market multiple derived from the average valuation of 42 hand-picked public companies. Some of the data set's highly relevant companies did not exist five years ago. Others will not exist five years from now. Company selection, therefore, must be adjusted over time as new companies IPO and others disappear due to M&A or bankruptcy. Additionally, each quarter, a few of the 42 companies will have ridiculous valuations. The 90x revenue multiple must be manually adjusted or removed because it does not represent reality, but some company-specific quirk related to M&A, revenue recognition, or a one-time accounting issue.

Maintaining such data sets becomes time intensive, and the prospect of subjective adjustments raises concerns about "cherry picking". Why are these 42 companies reasonable? How were spurious numbers dealt with? Has some data been added or excluded, and did that have a favorable or negative influence on the result? LPs or regulators may require explanations and backup for each decision.

The hassle of company selection can be eliminated by using the components of a basic market index or ETF to define a "market". This makes it someone else's job to decide whether companies should be added or removed. Many indices are designed explicitly to represent a "market", and because index values and components are public information, they are easy to look up, verify, and defend. Furthermore, market intelligence platforms like S&P CapitalIQ can usually simplify data processing for known indices with direct measurements of index valuation or P&L metrics.

Working with the S&P 500 Software Industry Index (^SP500-451030), the "Index" (S&P CapitalIQ 2020), will prove much easier than a hand-picked list of 42 software companies. It contains about a dozen companies including Microsoft, Adobe, Oracle, etc. Software Index values increased six-to-seven-fold over the decade. This may be viewed as the opportunity cost of not investing in large enterprise software stocks. Like PME, it can also be interpreted as the Market-driven Return of investments in the space. For example, since the Index value is $812 on 12/31/2013 and $1,642 on 12/31/2017, PME indicates that any investments in the space must generate a Gross MOIC of 2.0x and Gross IRR of 19.2% over those four years to beat the "market".

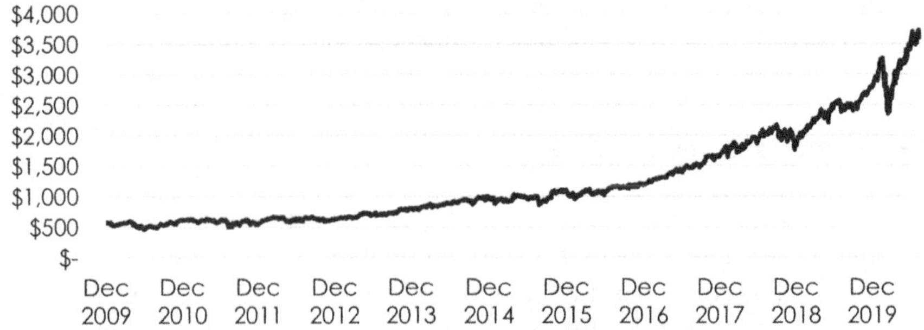

Figure 12.1 S&P 500 Software Industry index values

Index values are a good basis for PME, but the value creation models require market proxies for valuation and addressable market. These can be derived from the Index component companies, with functions of company valuation defining the market multiple and Index total revenue defining market growth. The use of Index total revenue is similar to Gross Output in the prior section. It is not assumed that Index total revenue is the same as the addressable market, but if the twelve largest software companies grow combined sales by 10%, it might be assumed that β_R for a company operating in that industry also grows by about 10%.

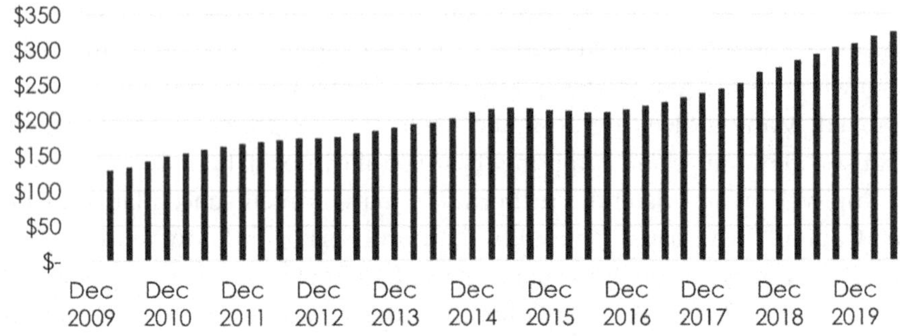

Figure 12.2 S&P 500 Software Industry index total TTM revenue

Index total revenue more than doubled over the decade, with values under $150 million in 2010 surpassing $300 million in recent quarters. Index total revenue grew from $189 million on 12/31/2013 to $238 million on 12/31/2017, providing a β_R of 25.8%. This suggests that value creation driven by Addressable Market Growth can be represented by an equity return multiplier of 1.26x and value creation IRR of 5.9%.

Market multiples for a software index are measured in the form of enterprise valuation divided by revenue. Valuation growth over the decade has been significant, with recent multiples surpassing 10.0x after troughing below 3.0x in 2011.

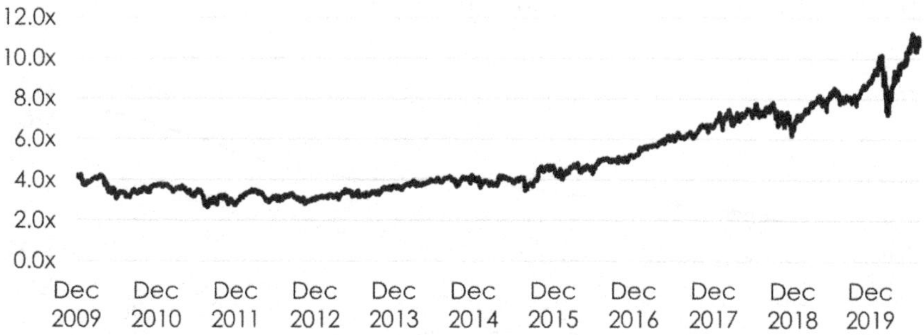

Figure 12.3 S&P 500 Software Industry index multiples

The Index multiples are quite volatile, which is a function of measurement and information availability. The Index multiple's denominator, revenue, is a trailing-twelve-month (TTM) value. It changes gradually because any one-day change is a small component of TTM. Furthermore, Index components rarely report revenue data more than once per quarter. In contrast, the Index multiple's numerator is enterprise valuation, which can change every minute as component share prices change.

Smoothing functions, like moving averages, are common in public equity analysis because they eliminate noise from price movements. It is reasonable to apply a similar function to Index multiples. Otherwise, share price-driven volatility injects noise into the value creation analysis, which can generate spurious results.

Daily valuation fluctuations have little bearing on private company investments that are valued only once per quarter. The same is true of transaction values determined at entry and exit. Purchase price is the outcome of a complicated due diligence process and intense price negotiation that may span three months, six months, or longer. It makes no sense for a six-month transaction process to have a materially different market multiple because the transaction closed on a Friday rather than a Tuesday when stock prices are 5% higher.

Figure 12.4 reduces Index multiples to their quarterly averages, or three-month simple moving average (SMA). The transformation tempers volatility and limits the data set's range. Daily Index multiples range from 2.7x to 10.7x, whereas quarterly Index multiples range from 3.0x to 9.5x.

Quarterly Index multiples grew from 3.5x on 12/31/2013 to 6.5x on 12/31/2017, representing an increase of 85%. This suggests that value creation driven by Market Multiple Expansion can be represented by an equity return multiplier of 1.85x and value creation IRR of 16.6%.

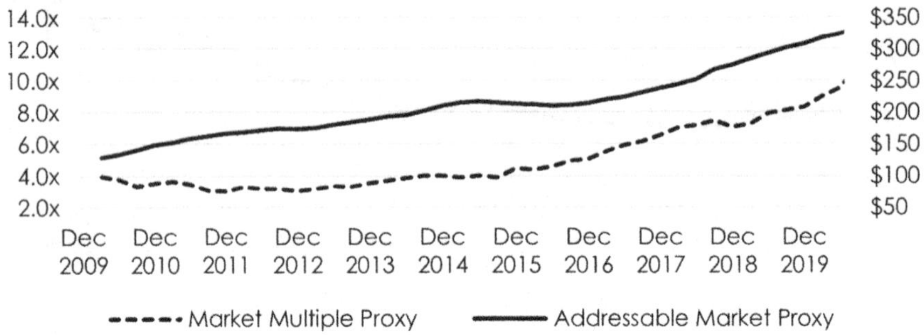

Figure 12.4 S&P 500 Software Industry index market proxies

Market Multiple Expansion and Addressable Market Growth combine to produce a Market-driven Return that is the product of 1.26x and 1.85x, or 2.33x. This suggests a Market-driven IRR of 23.5%. If a Fund investment in the space generates a 2.00x Gross MOIC and 18.9% Gross IRR, the market-adjusted, Manager-driven Return multiple would be 2.00x/2.33x, or 0.86x, and the Manager-driven IRR would be -3.7%, suggesting that the investment underperforms the market.

As demonstrated in the prior section, an annual Market Return Matrix can provide a useful snapshot of return expectations across various entry and exit date combinations.

Table 12.5 Annual software market proxies

	2010	2011	2012	2013	2014	2015	2016	2017	2018	2019	2020
Index Total Revenue	*$134*	*$159*	*$172*	*$181*	*$195*	*$217*	*$210*	*$225*	*$251*	*$294*	*$324*
Index Multiple	*3.7x*	*3.3x*	*3.2x*	*3.3x*	*3.9x*	*4.1x*	*4.7x*	*6.0x*	*7.2x*	*7.9x*	*9.5x*

Table 12.5 provides the necessary input variables, with Index total revenue set to mid-year (TTM June 30) and Index multiples set to the average of all daily Index multiples measured in each year. These numbers can be aggregated quarterly, or even monthly, but given the multitude of possible

combinations of entry and exit dates that can occur over 10 years, the annual Market Return Matrix snapshot is easiest to process:

Table 12.6 Software market Return Matrix by entry and exit year

		Exit or Current Year										
		2010	2011	2012	2013	2014	2015	2016	2017	2018	2019	2020
Entry Year	2010	*1.00x*	*1.07x*	*1.11x*	*1.23x*	*1.56x*	*1.80x*	*2.03x*	*2.76x*	*3.69x*	*4.75x*	*6.31x*
	2011	-	*1.00x*	*1.04x*	*1.15x*	*1.45x*	*1.68x*	*1.89x*	*2.58x*	*3.44x*	*4.43x*	*5.89x*
	2012	-	-	*1.00x*	*1.11x*	*1.40x*	*1.62x*	*1.83x*	*2.49x*	*3.32x*	*4.28x*	*5.68x*
	2013	-	-	-	*1.00x*	*1.26x*	*1.46x*	*1.65x*	*2.24x*	*2.99x*	*3.86x*	*5.12x*
	2014	-	-	-	-	*1.00x*	*1.16x*	*1.30x*	*1.77x*	*2.37x*	*3.05x*	*4.05x*
	2015	-	-	-	-	-	*1.00x*	*1.13x*	*1.53x*	*2.05x*	*2.64x*	*3.51x*
	2016	-	-	-	-	-	-	*1.00x*	*1.36x*	*1.82x*	*2.34x*	*3.11x*
	2017	-	-	-	-	-	-	-	*1.00x*	*1.34x*	*1.72x*	*2.29x*
	2018	-	-	-	-	-	-	-	-	*1.00x*	*1.29x*	*1.71x*
	2019	-	-	-	-	-	-	-	-	-	*1.00x*	*1.33x*
	2020	-	-	-	-	-	-	-	-	-	-	*1.00x*

Table 12.6 suggests that representative software deals should generate Gross MOICs of at least 1.56x if held from 2010 to 2014, 1.83x if held from 2012 to 2016, 2.37x if held from 2014 to 2018, and so on. The matrix provides a useful representation of Market-driven Returns that can fit onto an index card.

With performance details typically available in a Fund quarterly report or Annual Meeting deck, an analyst can use such a table to quickly estimate each investment's market-adjusted return in the form of MOIC, IRR, relative (%) value creation, etc. Calculations could not be described as "back of the envelope" because formulas with exponents and natural logarithms require a calculator. However, unlike PME, the analysis requires no time series data or complicated cash flow models. Manager-driven Returns could be estimated in "real time", for example, as an LP is attending an Annual GP Meeting.

Once an appropriate Industry Index is chosen, the Market Return Matrix data set is relatively easy to maintain. Outsourcing company selection to the professionals that design and established market indices for a living takes care of most of the work. Then, using market intelligence platforms like S&P CapitalIQ, data sets for a dozen different "markets" can be built or updated in a day.

12.7. Application: Continuous Change

The Market Return Matrices above sacrifice some precision for usability. Annual measures omit quarterly or monthly market movements, which can be material. Table 12.4 and Table 12.6 each provide about 50 data points.

Quarterly measurements would expand it to 800 data points. Monthly measurements would expand it to 7,200 data points. Such data sets are unsuitable in hard copy but could form the basis of a spreadsheet or software application that extracts appropriate β_R, MM_1, and MM_2 values based on investment entry and exit dates.

The special properties of the natural logarithm suggest an approach that is particularly well suited to automation. Derivations of the Logarithmic Model of value creation (Section 5.5 and Section 6.3) referred to continuous change (r_σ), where a change in σ between times t_1 and t_2 is expressed:

$$r_{1,2} = \ln(\sigma_2) - \ln(\sigma_1) = \ln\left(\frac{\sigma_2}{\sigma_1}\right)$$

For continuous change, $\sigma_2 = \sigma_1 \cdot e^{r_{1,2}}$. The expression has the same structure as formulas for continuous compounding interest rate and present/future value. This differs from the standard change (R_σ), where $\sigma_2 = \sigma_1 \cdot \left(1 + R_{1,2}\right)$, because:

$$R_{1,2} = \frac{\sigma_2 - \sigma_1}{\sigma_1} = \frac{\sigma_2}{\sigma_1} - 1$$

The continuous form has two useful properties. First, the magnitudes of $r_{i,j}$ values are independent of direction (increase vs. decrease). Second, the $r_{i,j}$ values are additive, meaning $r_{i,j}$ for a five-year hold is equal to the sum of the five annual $r_{i,j}$ values, 20 quarterly $r_{i,j}$ values, 60 monthly $r_{i,j}$ values, etc.

For example, consider a variable that starts at \$100 at t_1, increases to \$125, drops back to \$100, increases to \$125 again, and then finally returns to \$100 at t_5. Since σ_1 and σ_5 are both \$100, $r_{1,5}$ and $R_{1,5}$ must each be 0%. The standard and continuous changes for each step are as follows:

Table 12.7 Continuous and standard rates of change

	σ	$r_{i,j}$	$R_{i,j}$
t_1	\$100	-	-
t_2	\$125	+22.3%	+25.0%
t_3	\$100	-22.3%	-20.0%
t_4	\$125	+22.3%	+25.0%
t_5	\$100	-22.3%	-20.0%
Total		0.0%	10.0%

Unlike the standard change values, the continuous change values are all the same magnitude and sum to 0%. The four $r_{i,j}$ values sum to $r_{1,5}$ because:

$$r_{1,2} + r_{2,3} + r_{3,4} + r_{4,5} = \ln\left(\frac{\sigma_2}{\sigma_1}\right) + \ln\left(\frac{\sigma_3}{\sigma_2}\right) + \ln\left(\frac{\sigma_4}{\sigma_3}\right) + \ln\left(\frac{\sigma_5}{\sigma_4}\right)$$

$$= \ln\left(\frac{\sigma_2}{\sigma_1} \cdot \frac{\sigma_3}{\sigma_2} \cdot \frac{\sigma_4}{\sigma_3} \cdot \frac{\sigma_5}{\sigma_4}\right) = \ln\left(\frac{\sigma_5}{\sigma_1}\right) = r_{1,5}$$

Additive changes are useful when dealing with large data sets, like the input variables required to calculate Addressable Market Growth, Market Multiple Expansion, and Market-driven Returns. Table 12.8 provides the S&P 500 Software Industry Index (^SP500-451030) market proxies, with the quarterly continuous changes in Index total revenue describing r_{AMG} and Index valuation describing r_{MME}. The r_{MARKET} column is the sum of r_{AMG} and r_{MME}. This is the appropriate relationship because χ_{MARKET} is the product of χ_{AMG} and χ_{MME}, so the natural logarithms can be expressed:

$$\ln(\chi_{MARKET}) = \ln(\chi_{AMG} \cdot \chi_{MME}) = \ln(\chi_{AMG}) + \ln(\chi_{MME})$$

$$r_{MARKET} = r_{AMG} + r_{MME}$$

Table 12.8 Software market quarterly continuous changes

	Index Value	Index Total Revenue	Index Valuation	r_{AMG}	r_{MME}	r_{MARKET}
Mar 2010	$568	$130	3.99x	-	-	-
Jun 2010	$464	$134	3.77x	3.57%	-5.73%	-2.16%
Sep 2010	$535	$141	3.33x	5.16%	-12.41%	-7.25%
Dec 2010	$610	$149	3.55x	5.31%	6.39%	11.70%
Mar 2011	$606	$153	3.67x	2.79%	3.44%	6.23%
Jun 2011	$614	$159	3.47x	3.44%	-5.56%	-2.11%
Sep 2011	$545	$163	3.09x	2.47%	-11.76%	-9.29%
Dec 2011	$542	$166	3.03x	2.25%	-1.94%	0.30%
Mar 2012	$663	$169	3.29x	1.54%	8.28%	9.81%
Jun 2012	$636	$172	3.19x	1.66%	-3.12%	-1.46%
Sep 2012	$640	$174	3.19x	1.40%	-0.10%	1.31%
Dec 2012	$621	$174	3.02x	-0.20%	-5.46%	-5.66%
Mar 2013	$657	$176	3.19x	0.94%	5.55%	6.49%
Jun 2013	$701	$181	3.33x	2.98%	4.27%	7.24%
Sep 2013	$727	$184	3.31x	1.84%	-0.65%	1.19%
Dec 2013	$812	$189	3.53x	2.50%	6.68%	9.18%
Mar 2014	$865	$194	3.72x	2.64%	5.00%	7.64%
Jun 2014	$887	$195	3.86x	0.79%	3.75%	4.54%
Sep 2014	$929	$202	4.04x	3.45%	4.52%	7.97%
Dec 2014	$977	$211	4.02x	3.99%	-0.46%	3.53%
Mar 2015	$916	$215	3.90x	2.04%	-2.91%	-0.87%
Jun 2015	$953	$217	4.02x	0.98%	3.09%	4.07%
Sep 2015	$925	$216	3.90x	-0.65%	-3.23%	-3.88%
Dec 2015	$1,086	$213	4.44x	-1.08%	13.06%	11.99%
Mar 2016	$1,097	$212	4.39x	-0.43%	-1.21%	-1.64%

	Index Value	Index Total Revenue	Index Valuation	r_{AMG}	r_{MME}	r_{MARKET}
Jun 2016	$1,071	$210	4.59x	-1.13%	4.47%	3.34%
Sep 2016	$1,156	$211	4.93x	0.45%	7.11%	7.56%
Dec 2016	$1,184	$214	5.04x	1.59%	2.40%	3.99%
Mar 2017	$1,318	$220	5.51x	2.51%	8.86%	11.38%
Jun 2017	$1,412	$225	5.92x	2.12%	7.19%	9.31%
Sep 2017	$1,503	$231	6.16x	3.04%	4.00%	7.04%
Dec 2017	$1,642	$238	6.53x	2.66%	5.82%	8.49%
Mar 2018	$1,764	$244	7.03x	2.54%	7.28%	9.82%
Jun 2018	$1,904	$251	7.20x	3.10%	2.45%	5.55%
Sep 2018	$2,161	$268	7.47x	6.39%	3.63%	10.02%
Dec 2018	$1,909	$275	7.08x	2.42%	-5.32%	-2.90%
Mar 2019	$2,242	$285	7.28x	3.64%	2.76%	6.39%
Jun 2019	$2,458	$294	7.95x	3.07%	8.83%	11.90%
Sep 2019	$2,489	$303	8.15x	3.04%	2.45%	5.49%
Dec 2019	$2,781	$308	8.35x	1.81%	2.44%	4.25%
Mar 2020	$2,708	$318	9.02x	3.26%	7.73%	10.99%
Jun 2020	$3,518	$324	9.53x	1.87%	5.53%	7.39%

Table 12.8 indicates that from 12/31/2013 to 12/31/2017 the Index total revenue grows from $189 billion to $238 billion, while Index valuation grows from 3.5x to 6.5x. This provides an r_{AMG} of ln($238/$189), or 23.0%, and an r_{MME} of $\ln(6.5x/3.5x)$, or 61.5%. The sum of r_{AMG} and r_{MME} provide an r_{MARKET} value of 84.5%. These holding period continuous change values are identical to the sum of all quarterly continuous changes that occur over the hold. Adding up the 16 rows starting at Mar 2014 and ending at Dec 2017 provides 23.0% in the r_{AMG} column, 61.5% in the r_{MME} column, and 84.5% in the r_{MARKET} column.

To generalize this relationship, the total continuous change in σ is defined as r_σ. The various intermediate continuous changes in σ that occur over the holding period are defined as $r_{\sigma(i)}$. Since r_σ is equal to the sum of all $r_{\sigma(i)}$ values:

$$r_\sigma = \sum_i r_{\sigma(i)}$$

Therefore, the relationship between the holding period r_{MARKET} value and the intermediate continuous changes can be expressed:

$$r_{MARKET} = \sum_i r_{MARKET(i)} = \sum_i r_{AMG(i)} + \sum_i r_{MME(i)}$$

$$= r_{MARKET(Q1)} + r_{MARKET(Q2)} + r_{MARKET(Q3)} + \cdots$$

$$= r_{AMG(Q1)} + r_{MME(Q1)} + r_{AMG(Q2)} + r_{MME(Q2)} + r_{AMG(Q3)} + r_{MME(Q3)} + \cdots$$

This expression may be substituted into the value creation component formulas. For example, market-driven relative (%) value creation can be expressed:

$$v_{MARKET} = \frac{\ln(\chi_{MARKET})}{\ln(MOIC)}$$

$$= \frac{\sum_i r_{MARKET(i)}}{\ln(MOIC)} = \frac{\sum_i r_{AMG(i)} + \sum_i r_{MME(i)}}{\ln(MOIC)}$$

If the Fund investment has a Gross MOIC of 4.0x, then market-driven relative (%) value creation is:

$$v_{MARKET} = \frac{\sum_i r_{MARKET(i)}}{\ln(MOIC)} = \frac{84.42\%}{\ln(4.0x)} = 60.9\%$$

What advantage do these Σ-formulations offer? Plugging $\ln(MM_2/MM_1)$ into a handheld calculator is certainly easier than adding up 16 quarterly continuous change values. The Σ-formulations are helpful when a "standard" data set is used to evaluate many different investments, with various entry and exit dates. In such cases, calculations of Addressable Market Growth, Market Multiple Expansion, and Market-driven Returns can be automated using conditional functions. For example, the function might sum all $r_{MME(i)}$ values that are greater than the entry date but less than or equal to the exit date, and then use the resulting $\sum_i r_{MME(i)}$ value directly in the value creation formulas.

In the Σ-formulation, Market-driven absolute ($) value creation can be determined from relative (%) value creation:

$$V_{MARKET} = \Delta FEqV \cdot v_{MARKET}$$

$$= \Delta FEqV \cdot \frac{\sum_i r_{MARKET(i)}}{\ln(MOIC)} = \Delta FEqV \cdot \frac{\sum_i r_{AMG(i)} + \sum_i r_{MME(i)}}{\ln(MOIC)}$$

If r_{MARKET} is the natural logarithm of χ_{MARKET}, then χ_{MARKET} must be equal to e raised to the power of r_{MARKET}. Therefore, in the Σ-formulation:

$$\chi_{MARKET} = e^{r_{MARKET}} = e^{\sum_i r_{MARKET(i)}} = e^{\left(\sum_i r_{AMG(i)} + \sum_i r_{MME(i)}\right)}$$

$$= e^{\sum_i r_{AMG(i)}} \cdot e^{\sum_i r_{MME(i)}}$$

$$= \chi_{AMG} \cdot \chi_{MME}$$

Market-driven IRRs can be determined from equity return multipliers. Therefore, in the Σ-formulation:

$$IRR_{MARKET} = \chi_{MARKET}{}^{1/\Delta t} - 1$$

$$= \left[e^{\sum_i r_{MARKET(i)}}\right]^{1/\Delta t} - 1 = \left[e^{\left(\sum_i r_{AMG(i)} + \sum_i r_{MME(i)}\right)}\right]^{1/\Delta t} - 1$$

$$= e^{\sum_i r_{MARKET(i)}/\Delta t} - 1 = e^{\left(\sum_i r_{AMG(i)} + \sum_i r_{MME(i)}\right)/\Delta t} - 1$$

The Σ-formulations are mathematically identical to the Logarithmic Model Market-driven Returns described in Section 12.2, and they can be used to determine the Manager-driven Returns described in Section 12.3. Using these Σ-formulations, Logarithmic Model value creation can be expressed:

$$\Delta FEqV = V_{MARKET} + V_{MANAGER}$$

$$= \Delta FEqV \frac{\sum r_{MARKET}}{\ln(MOIC)} + \Delta FEqV \left(1 - \frac{\sum r_{MARKET}}{\ln(MOIC)}\right)$$

12.8. Summary

The three value creation models provide complete Fund-level descriptions of Market-driven and Manager-driven Returns. Each model can express returns in the form of absolute ($) value creation, relative (%) value creation, equity return multiples or multipliers, and value creation IRRs. As expected, results provided by the Derivative and Logarithmic models are nearly identical. The Market-driven return is typically slightly higher in the Conventional Model because of its tendency to understate the impact of leverage.

The significance of the analysis is limited by the accuracy and relevance of the underlying proxies for market multiples and addressable markets. This is true of any measurement of market-adjusted returns, including PME. The examples in Section 12.5 and Section 12.6 suggest a framework and some ideas for building proxies for various markets or industries based on private company M&A data, economic data, and public company indices.

Since value creation is driven by relative (%) changes in market proxies, rather than proxy absolute ($) values, they need not be perfect. It is usually sufficient for proxies to move in the right direction and exhibit changes of reasonable magnitude. It is important that proxies be non-volatile, but otherwise, it is generally better for proxies to be transparent, uncontroversial, and easy to update than it is for them to be precise, especially if precision comes with the cost of dealing with complicated models.

The next chapter investigates the impact of holding period events, like add-on acquisitions and dividend recapitalizations, on the value creation analysis.

12.9. Optional: An Acceptable Indirect Value Driver Calculation

The Manager-driven Return formula in Section 12.3 breaks the rule outlined in Section 3.4, that all value drivers should be calculated "directly". The rule helps to prevent errors and biases. For example, knowing that $\Delta FEqV$ may be taken from Fund reporting materials, an analyst may attempt to calculate V_{EG} and V_{ME} directly and then solve V_{LEV} indirectly, as follows:

$$V_{LEV} = \Delta FEqV - V_{EG} - V_{ME}$$

With the indirect approach above, any errors or biases in the calculation of V_{EG} or V_{ME} would accumulate in V_{LEV}.

Now, consider the expression for Manager-driven absolute ($) value creation:

$$V_{MANAGER} = \Delta FEqV - V_{MARKET} = \Delta FEqV - V_{AMG} - V_{MME}$$

Not only is this an indirect calculation, but the V_{AMG} and V_{MME} values most certainly have errors because they are calculated from imperfect market proxies. However, this formula always provides the same result as the direct version of the $V_{MANAGER}$ calculation, which includes the components V_{MSG}, V_{EME}, V_{IME}, V_{LEV}, and V_{FO} and depends on several P&L and balance sheet input variables. How is this possible?

In this case, the shortcut works because V_{MSG} and V_{IME} correct for any errors in V_{AMG} and V_{MME}. Revenue Growth is a combination of Addressable Market Growth and Market Share Growth, so any errors in V_{AMG} are offset by V_{MSG} and, therefore, have no impact on V_{RG}. Similarly, Multiple Expansion is a combination of Market Multiple Expansion and Intrinsic Multiple Expansion, so any errors in V_{MME} are offset by V_{IME} and, therefore, have no impact on V_{ME}. The same offsets occur in the other value creation measures: relative (%) value creation, equity return multipliers, and value creation IRRs.

It would be best to ensure that the proxies for market multiples and addressable markets are accurate and relevant, but it is good to know that if they are not, those errors are "contained" and have no impact on the core capital structure and P&L-driven value creation components, including V_{FO}, V_{ISO}, V_{UNLEV}, $V_{CF/LEV}$, V_{EG}, V_{ME}, V_{RG}, V_{EME}, V_{GME}, and V_{OMR}.

13. Holding Period Events

This chapter describes the impact of holding period events on value creation analysis. Inflows, like follow-on equity investments and add-on acquisitions, change entry (t_1) values to effective entry (t_1') values. Outflows, like dividend recapitalizations and divestments, change exit (t_2) values to effective exit (t_2') values. Input adjustments can be "direct" or "indirect". Adding incremental EBITDA from an add-on to E_1 is direct. Recalculating M_1 based on adjustments to EBITDA and enterprise valuation is indirect. Required value creation input adjustments for several holding period events are as follows:

Table 13.1 Required direct and indirect adjustments to value creation input variables

t_1	Equity Investment	Add-on Investment	t_2	Dividend Recapitalization	Divestment
$FEqV_1$	Direct	Direct	$FEqV_2$	Direct	Direct
$TEqV_1$	Direct	Direct	$TEqV_2$	Direct	Direct
FO_1	Via $FEqV, TEqV$	Via $FEqV, TEqV$	FO_2	Via $FEqV, TEqV$	Via $FEqV, TEqV$
TEV_1	Via $TEqV$	Via $TEqV, ND$	TEV_2	-	Via $TEqV, ND$
ND_1	-	Direct	ND_2	Direct	Direct
ER_1	Via $TEqV$	Via $TEqV, TEV$	ER_2	Via $TEqV$	Via $TEqV, TEV$
E_1	-	Direct	E_2	-	Direct
M_1	Via TEV	Via TEV, E	M_2	-	Via TEV, E
R_1	-	Direct	R_2	-	Direct
EM_1	-	Via R, E	EM_2	-	Via R, E
GP_1	-	Direct	GP_2	-	Direct
GM_1	-	Via GP, R	GM_2	-	Via GP, R
OM_1	-	Via GM, EM	OM_2	-	Via GM, EM
MM_1	-	-	MM_2	-	-
μ_1	Via M	Via M	μ_2	-	Via M
β_R	-	-	β_R	-	-
α_R	-	Via R	α_R	-	Via R

13.1. Background

What about add-ons and dividend recaps? Are they value drivers? How do they impact value creation? This chapter evaluates these and other events that result in capital flowing into or out of private company investments during a holding period.

This is the last chapter on drivers of Fund-level Gross Returns. Prior chapters described three different models (Conventional, Derivative, and Logarithmic) for calculating 16 different value creation components (abbreviated *FO, ISO, UNLEV, CF/LEV, EG, ME, RG, EME, GME, OMR, AMG, MSG, MME, IME, MARKET*, and *MANAGER*) that can be measured in four different ways (absolute ($) value creation, relative (%) value creation, equity return multiples or multipliers, and value creation IRRs). So far, that is about 192 return and value creation formulas.

Events like add-ons and dividend recaps generally are not value drivers, but they can impact every other value creation component because they shift the effective starting point or ending point of the analysis. If a Fund makes an initial investment of $5 at t_1, a follow-on investment of $5, and then realizes $20 at t_2, its $\Delta FEqV$ is $10 (not $15) and Gross MOIC is 2.0x (not 4.0x) because follow-on investments during the hold reduce returns.

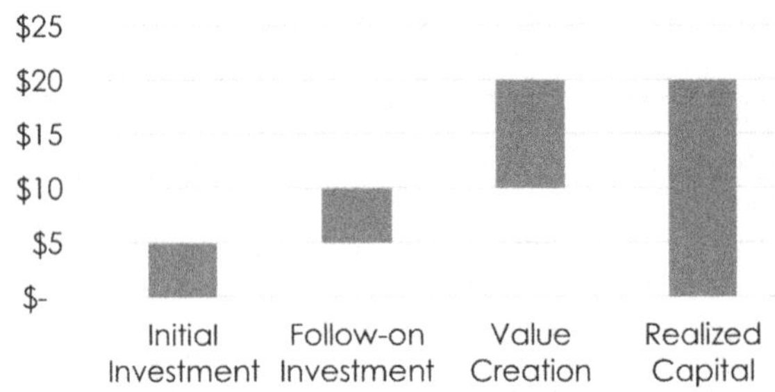

Figure 13.1 Value creation bridge with follow-on investment

The follow-on investment may look like a value driver in Figure 13.1, but it actually reduces value creation because the analysis starting point is the sum of the first two columns. All 16 value creation components must fit into the third column, "bridging the gap" between $10 and $20. Similarly, if a Fund invests $10 at t_1, distributes $10 during the hold, and then realizes $20 at t_2, its $\Delta FEqV$ is $20 (not $10) and Gross MOIC is 3.0x (not 2.0x) because distributions during the hold increase returns.

Adjusting the return expressions to include inflows ($FEqV_{in}$) like follow-on investments and outflows ($FEqV_{out}$) like distributions provides:

$$\Delta FEqV = FEqV_2 + FEqV_{out} - FEqV_1 - FEqV_{in} = FEqV_2' - FEqV_1'$$

$$MOIC = \frac{FEqV_2 + FEqV_{out}}{FEqV_1 + FEqV_{in}} = \frac{FEqV_2'}{FEqV_1'}$$

The superscript ($'$) identifies an "effective" input variable, where adjustments have been made for all appropriate inflows and outflows, so $FEqV_1'$ is equivalent to $FEqV_1 + FEqV_{in}$, and $FEqV_2'$ is equivalent to $FEqV_2 + FEqV_{out}$.

Most of the other value creation input variables require similar adjustments. If a company has EBITDA of \$5 at t_1, acquires additional EBITDA of \$5 through an add-on acquisition, and then has EBITDA of \$20 at t_2, its ΔE is \$10 (not \$15) and E_2/E_1 is 2.0x (not 4.0x) because acquired EBITDA reduces EBITDA Growth. Adjusting the EBITDA change expressions to include inflows (E_{in}) and outflows (E_{out}) provides:

$$\Delta E = E_2 + E_{out} - E_1 - E_{in} = E_2' - E_1'$$

$$\chi_{EG} = \frac{E_2 + E_{out}}{E_1 + E_{in}} = \frac{E_2'}{E_1'}$$

E_{in} represents acquired EBITDA that did not exist at t_1 and, therefore, reduces EBITDA Growth. E_{out} represents divested EBITDA that is excluded from the company P&L at t_2 and, therefore, increases EBITDA Growth. Events including E_{out} values are less common, but they can occur if a company sells off a subsidiary or business unit.

The *FEqV* and *E* adjustments are "direct" because they are simply added to or subtracted from t_1 or t_2 values. In contrast, "indirect" adjustments are the result of direct adjustments to other input variables. For example, adjustments to valuation multiples are indirectly driven by adjustments to *TEV* and *E*, so the valuation multiple change expressions are:

$$\Delta M = \frac{TEV_2 + TEV_{out}}{E_2 + E_{out}} - \frac{TEV_1 + TEV_{in}}{E_1 + E_{in}} = \frac{TEV_2'}{E_2'} - \frac{TEV_1'}{E_1'} = M_2' - M_1'$$

$$\chi_{ME} = \frac{(TEV_2 + TEV_{out})/(E_2 + E_{out})}{(TEV_1 + TEV_{in})/(E_1 + E_{in})} = \frac{TEV_2'/E_2'}{TEV_1'/E_1'} = \frac{M_2'}{M_1'}$$

The adjusted value creation input variables are referred to as "effective" values, so if M_1 is the entry multiple, M_1' would be the effective entry multiple, which is fully adjusted for EBITDA and enterprise valuation acquired after the initial investment.

The following sections describe six holding period events and examine how they impact every value creation component. The direction of change is noted for each input variable, along with an indication of whether they have a positive (↑), negative (↓), or indeterminate (±) impact on value creation components. Because each event causes numerous changes to input variables, a systematic approach is required to ensure that all adjustments are made, and the 192 value creation component formulas described thus far continue to work as intended.

13.2. Follow-on Equity Investments

Capital required to support growth initiatives, or fend off lenders, is usually provided by the Fund, co-investors, or both. Follow-on equity investments impact company capital structures, but not P&Ls. Since equity investments are inflows, adjustments are made to t_1 values:

- Fund equity investments directly increase $FEqV_1'$ ($\downarrow \Delta FEqV$).
- Fund and co-investor investments directly increase $TEqV_1'$ ($\downarrow V_{ISO}$).
- Increases in $FEqV_1'$ and $TEqV_1'$ indirectly change FO_1' ($\pm V_{FO}$).
- Increases in $TEqV_1'$ indirectly increase TEV_1' ($\downarrow V_{UNLEV}$).
- Increases in $TEqV_1'$ and TEV_1' indirectly change ER_1' ($\pm V_{LEV}$).
- Increases in TEV_1' indirectly increase M_1' ($\downarrow V_{ME}$).
- Increases in M_1' indirectly increase μ_1' ($\downarrow V_{IME}$).

Consider an investment in CharlieCo, just like AlphaCo in prior chapters, but where the Fund and co-investors each provide \$5 of follow-on equity capital. The effective value creation input variables would be as follows:

Table 13.2 CharlieCo effective value creation input variables after follow-on equity

	t_1	t_{in}	t'_1	t_2	ΔX	$\overline{X}$	χ_X
FEqV	*\$44*	*+\$5*	*\$49*	*\$88*	*\$39*	-	*1.80x*
TEqV	*\$80*	*+\$10*	*\$90*	*\$180*	*\$90*	*\$135*	*2.00x*
FO	*55%*	*Indirect*	*54.4%*	*48.9%*	*-5.6%*	*51.7%*	*0.90x*
ND	*\$20*	-	*\$20*	*\$0*	*- \$20*	-	-
TEV	*\$100*	*Indirect*	*\$110*	*\$180*	*\$70*	*\$145*	*1.64x*
ER	*80%*	*Indirect*	*81.8%*	*100%*	*18.2%*	*90.9%*	*1.22x*
E	*\$20*	-	*\$20*	*\$30*	*\$10*	*\$25*	*1.50x*
M	*5.0x*	*Indirect*	*5.5x*	*6.0x*	*0.5x*	*5.75x*	*1.09x*
R	*\$80*	-	*\$80*	*\$100*	*\$20*	*\$90*	*1.25x*
EM	*25%*	-	*25%*	*30%*	*5%*	*27.5%*	*1.20x*
GP	*\$48*	-	*\$48*	*\$62*	-	-	-
GM	*60%*	-	*60%*	*62%*	*2%*	-	*1.08x*
OM	*35%*	-	*35%*	*32%*	*-3%*	-	*1.12x*
MM	*7.0x*	-	*7.0x*	*8.0x*	*1.0x*	*7.5x*	*1.14x*
μ	*71.4%*	*Indirect*	*78.6%*	*75.0%*	*-3.6%*	*76.8%*	*0.95x*
β_R	-	-	-	*10.0%*	-	-	*1.10x*
α_R	-	-	-	*13.6%*	-	-	*1.14x*

These effective (t'_1) value creation input variables provide the measures of ΔX and $\overline{X}$ needed for the Conventional and Derivative models and the measures of χ_X required for the Logarithmic Model.

13.3. Subsequent Debt Investments

Capital needs may also be met by raising debt capital. Subsequent debt investments impact company capital structures, but not P&Ls. Adjusting for subsequent debt investments ensures that companies get appropriate credit for cash flow generation or delevering. Since debt investments are inflows, adjustments are made to t_1 values:

- Company debt raises directly increase ND'_1 ($\uparrow V_{CF}$).
- Increases in ND'_1 indirectly increase TEV'_1 ($\downarrow V_{UNLEV}$).
- Increases in TEV'_1 indirectly decrease ER'_1 ($\uparrow V_{LEV}$) and increase M'_1 ($\downarrow V_{ME}$).
- Increases in M'_1 indirectly increase μ'_1 ($\downarrow V_{IME}$).

Consider an investment in DeltaCo, just like AlphaCo in prior chapters, but with \$20 of additional debt issued during the holding period. The effective value creation input variables would be as follows:

Table 13.3 DeltaCo effective value creation input variables after subsequent debt

	t_1	t_{in}	t_1'	t_2	ΔX	$\bar{X}$	χ_X
FEqV	*$44*	-	*$44*	*$88*	*$44*	-	*2.00x*
TEqV	*$80*	-	*$80*	*$180*	*$100*	*$130*	*2.25x*
FO	*55%*	-	*55%*	*48.9%*	*-6.1%*	*51.9%*	*0.89x*
ND	*$20*	*+$20*	*$40*	*$0*	*-$40*		-
TEV	*$100*	*Indirect*	*$120*	*$180*	*$60*	*$150*	*1.50x*
ER	*80%*	*Indirect*	*66.7%*	*100%*	*33.3%*	*83.3%*	*1.50x*
E	*$20*	-	*$20*	*$30*	*$10*	*$25*	*1.50x*
M	*5.0x*	*Indirect*	*6.0x*	*6.0x*	*0.0x*	*6.00x*	*1.00x*
R	*$80*	-	*$80*	*$100*	*$20*	*$90*	*1.25x*
EM	*25%*	-	*25%*	*30%*	*5%*	*27.5%*	*1.20x*
GP	*$48*	-	*$48*	*$62*	-	-	-
GM	*60%*	-	*60%*	*62%*	*2%*	-	*1.08x*
OM	*35%*	-	*35%*	*32%*	*-3%*	-	*1.12x*
MM	*7.0x*	-	*7.0x*	*8.0x*	*1.0x*	*7.5x*	*1.14x*
μ	*71.4%*	*Indirect*	*85.7%*	*75.0%*	*-10.7%*	*80.4%*	*0.88x*
β_R	-	-	-	*10.0%*	-	-	*1.10x*
α_R	-	-	-	*13.6%*	-	-	*1.14x*

These effective (t_1') value creation input variables provide the measures of ΔX and $\bar{X}$ needed for the Conventional and Derivative models and the measures of χ_X required for the Logarithmic Model.

13.4. Add-on Acquisitions

Add-on acquisitions impact both capital structures and P&Ls of companies. They can be funded with equity, debt, or a combination of both. Since add-on acquisitions are inflows, adjustments are made to t_1 values:

- Fund equity investments directly <u>increase</u> $FEqV_1'$ ($\downarrow\Delta FEqV$).
- Fund and co-investor investments directly <u>increase</u> $TEqV_1'$ ($\downarrow V_{ISO}$).
- Company debt raises directly <u>increase</u> ND_1' ($\uparrow V_{CF}$).
- Increases in $FEqV_1'$ and $TEqV_1'$ indirectly <u>change</u> FO_1' ($\pm V_{FO}$).
- Increases in $TEqV_1'$ and ND_1' indirectly <u>increase</u> TEV_1' ($\downarrow V_{UNLEV}$).
- Increases in $TEqV_1'$ and TEV_1' indirectly <u>change</u> ER_1' ($\pm V_{LEV}$).
- Acquired EBITDA, gross profit, and revenue directly <u>increase</u> E_1', GP_1', and R_1' and indirectly <u>change</u> EM_1', GM_1', OM_1', and α_R (various impacts).
- Changes in TEV_1' and E_1' indirectly <u>change</u> M_1' ($\pm V_{ME}$).
- Changes in M_1' indirectly <u>change</u> μ_1' ($\pm V_{IME}$).

Consider an investment in EchoCo, just like AlphaCo in prior chapters, but with an add-on acquisition valued at $20. The transaction is funded with $5 from the Fund, $5 from co-investors, and $10 of debt. The add-on contributes $10 of revenue, $6 of gross profit, and $2.5 of EBITDA. The effective value creation input variables would be as follows:

Table 13.4 EchoCo effective value creation input variables after add-on

	t_1	t_{in}	t_1'	t_2	ΔX	$\overline{X}$	χ_X
FEqV	*$44*	*+$5*	*$49*	*$88*	*$39*	-	*1.80x*
TEqV	*$80*	*+$10*	*$90*	*$180*	*$90*	*$135*	*2.00x*
FO	*55%*	*Indirect*	*54.4%*	*48.9%*	*-5.6%*	*51.7%*	*0.90x*
ND	*$20*	*+$10*	*$30*	*$0*	*-$30*	-	-
TEV	*$100*	*Indirect*	*$120*	*$180*	*$60*	*$150*	*1.50x*
ER	*80%*	*Indirect*	*75%*	*100%*	*25.0%*	*87.5%*	*1.33x*
E	*$20*	*+$2.5*	*$22.5*	*$30*	*$7.5*	*$26*	*1.33x*
M	*5.0x*	*Indirect*	*5.3x*	*6.0x*	*0.7x*	*5.7x*	*1.13x*
R	*$80*	*+$10*	*$90*	*$100*	*$10*	*$95*	*1.11x*
EM	*25%*	*Indirect*	*25%*	*30%*	*5%*	*27.5%*	*1.20x*
GP	*$48*	*+$6*	*$54*	*$62*	-	-	-
GM	*60%*	*Indirect*	*60%*	*62%*	*2%*	-	*1.08x*
OM	*35%*	*Indirect*	*35%*	*32%*	*-3%*	-	*1.12x*
MM	*7.0x*	-	*7.0x*	*8.0x*	*1.0x*	*7.5x*	*1.14x*
μ	*71.4%*	*Indirect*	*76.2%*	*75.0%*	*-1.2%*	*75.6%*	*0.98x*
β_R	-	-	-	*10.0%*	-	-	*1.10x*
α_R	-	*Indirect*	-	*1.0%*	-	-	*1.01x*

These effective (t_1') value creation input variables provide the measures of ΔX and $\overline{X}$ needed for the Conventional and Derivative models and the measures of χ_X required for the Logarithmic Model.

13.5. Equity Distributions

Equity proceeds distributed to shareholders prior to exit impact company capital structures, but not P&Ls. They are effectively the opposite of follow-on equity investments. Since equity distributions are outflows, adjustments are made to t_2 values:

- Fund equity proceeds directly <u>increase</u> $FEqV_2'$ ($\uparrow \Delta FEqV$).
- Fund and co-investor proceeds directly <u>increase</u> $TEqV_2'$ ($\uparrow V_{ISO}$).
- Increases in $FEqV_2'$ and $TEqV_2'$ indirectly <u>change</u> FO_2' ($\pm V_{FO}$).
- Increases in $TEqV_2'$ indirectly <u>increase</u> TEV_2' ($\uparrow V_{UNLEV}$).
- Increases in $TEqV_2'$ and TEV_2' indirectly <u>change</u> ER_2' ($\pm V_{LEV}$).

- Increases in TEV_2' indirectly <u>increase</u> M_2' ($\uparrow V_{ME}$).
- Increases in M_2' indirectly <u>increase</u> μ_2' ($\uparrow V_{IME}$).

Consider an investment in FoxtrotCo, just like AlphaCo in prior chapters, but with an equity distribution of \$10 that provides proceeds of \$5 to the Fund and \$5 to co-investors. The effective value creation input variables would be as follows:

Table 13.5 FoxtrotCo effective value creation input variables after distribution

	t_1	t_2	t_{out}	t_2'	ΔX	$\bar{X}$	χ_X
FEqV	*\$44*	*\$88*	*+\$5*	*\$93*	*\$49*	-	*2.11x*
TEqV	*\$80*	*\$180*	*+\$10*	*\$190*	*\$110*	*\$135*	*2.38x*
FO	*55%*	*48.9%*	*Indirect*	*48.9%*	*-6.1%*	*52.0%*	*0.89x*
ND	*\$20*	*\$0*	-	*\$0*	*-\$20*	-	-
TEV	*\$100*	*\$180*	*Indirect*	*\$190*	*\$90*	*\$145*	*1.90x*
ER	*80%*	*100%*	*Indirect*	*100%*	*20.0%*	*90.0%*	*1.25x*
E	*\$20*	*\$30*	-	*\$30.0*	*\$10*	*\$25*	*1.50x*
M	*5.0x*	*6.0x*	*Indirect*	*6.3x*	*1.3x*	*5.7x*	*1.27x*
R	*\$80*	*\$100*	-	*\$100*	*\$20*	*\$90*	*1.25x*
EM	*25%*	*30%*	-	*30%*	*5%*	*27.5%*	*1.20x*
GP	*\$48*	*\$62*	-	*\$62*	-	-	-
GM	*60%*	*62%*	-	*62%*	*2%*	-	*1.08x*
OM	*35%*	*32%*	-	*32%*	*-3%*	-	*1.12x*
MM	*7.0x*	*8.0x*	-	*8.0x*	*1.0x*	*7.5x*	*1.14x*
μ	*71.4%*	*75.0%*	*Indirect*	*79.2%*	*7.7%*	*75.3%*	*1.11x*
β_R	-	*10.0%*	-	*10.0%*	-	-	*1.10x*
α_R	-	*13.6%*	-	*13.6%*	-	-	*1.14x*

These effective (t_2') value creation input variables provide the measures of ΔX and $\bar{X}$ needed for the Conventional and Derivative models and the measures of χ_X required for the Logarithmic Model.

13.6. Dividend Recapitalizations

Dividend recapitalizations are combinations of equity distributions and subsequent debt investments, where a company issues debt to provide equity proceeds to shareholders. They impact company capital structures, but not enterprise valuations or P&Ls. The positive $TEqV_{out}$ value is offset by a negative ND_{out} value of the same magnitude, which holds TEV_2' constant, increases $TEqV_2'$ to account for equity realized before t_2, and decreases ND_2' to account for additional debt paid down before t_2. Since equity distributions are outflows, adjustments are made to t_2 values:

- Fund equity proceeds directly <u>increase</u> $FEqV_2'$ ($\uparrow\Delta FEqV$).
- Fund and co-investor proceeds directly <u>increase</u> $TEqV_2'$ ($\uparrow V_{ISO}$).
- Increases in $FEqV_2'$ and $TEqV_2'$ indirectly <u>change</u> FO_2' ($\pm V_{FO}$).
- Debts incurred to provide proceeds directly <u>decrease</u> ND_2' ($\uparrow V_{CF}$).
- Increases in $TEqV_2'$ indirectly <u>change</u> ER_2' ($\pm V_{LEV}$).

Consider an investment in GolfCo, just like AlphaCo in prior chapters, but with a dividend recapitalization that adds $20 in debt to the company balance sheet while distributing proceeds of $10 to the Fund and $10 to co-investors. The effective value creation input variables would be as follows:

Table 13.6 GolfCo effective value creation input variables after dividend recap

	t_1	t_2	t_{out}	t_2'	ΔX	$\overline{X}$	χ_X
FEqV	*$44*	*$88*	*+$10*	*$98*	*$54*	-	*2.23x*
TEqV	*$80*	*$180*	*+$20*	*$200*	*$120*	*$140*	*2.50x*
FO	*55%*	*48.9%*	*Indirect*	*49.0%*	*-6.0%*	*52.0%*	*0.89x*
ND	*$20*	*$0*	*-$20*	*-$20*	*-$40*	-	-
TEV	*$100*	*$180*	-	*$180*	*$80*	*$140*	*1.80x*
ER	*80%*	*100%*	*Indirect*	*111%*	*31.1%*	*95.6%*	*1.39x*
E	*$20*	*$30*	-	*$30*	*$10*	*$25*	*1.50x*
M	*5.0x*	*6.0x*	-	*6.0x*	*1.0x*	*5.50x*	*1.20x*
R	*$80*	*$100*	-	*$100*	*$20*	*$90*	*1.25x*
EM	*25%*	*30%*	-	*30%*	*5%*	*27.5%*	*1.20x*
GP	*$48*	*$62*	-	*$62*	-	-	-
GM	*60%*	*62%*	-	*62%*	*2%*	-	*1.08x*
OM	*35%*	*32%*	-	*32%*	*-3%*	-	*1.12x*
MM	*7.0x*	*8.0x*	-	*8.0x*	*1.0x*	*7.5x*	*1.14x*
μ	*71.4%*	*75.0%*	-	*75.0%*	*3.6%*	*73.2%*	*1.05x*
β_R	-	*10.0%*	-	*10.0%*	-	-	*1.10x*
α_R	-	*13.6%*	-	*13.6%*	-	-	*1.14x*

These effective (t_2') value creation input variables provide the measures of ΔX and $\overline{X}$ needed for the Conventional and Derivative models and the measures of χ_X required for the Logarithmic Model.

13.7. Divestments

A divestment, like the sale of a company business unit, is the opposite of an add-on acquisition. They impact both capital structures and P&Ls of companies. In the sale, shareholders can receive consideration in the form of cash, equity, or the buyer's assumption of company liabilities. Since divestments are outflows, adjustments are made to t_2 values:

- Fund equity proceeds directly <u>increase</u> $FEqV_2'$ ($\uparrow\Delta FEqV$).
- Fund and co-investor proceeds directly <u>increase</u> $TEqV_2'$ ($\uparrow V_{ISO}$).
- Liabilities assumed by buyers directly <u>increase</u> ND_2' ($\downarrow V_{CF}$).
- Increases in $FEqV_2'$ and $TEqV_2'$ indirectly <u>change</u> FO_2' ($\pm V_{FO}$).
- Increases in $TEqV_2'$ and ND_2' indirectly <u>increase</u> TEV_2' ($\uparrow V_{UNLEV}$).
- Increases in $TEqV_2'$ and TEV_2' indirectly <u>change</u> ER_2' ($\pm V_{LEV}$).
- Divested EBITDA, gross profit, and revenue directly <u>increase</u> E_2', GP_2', and R_2' and indirectly <u>change</u> EM_2', GM_2', OM_2', and α_R (various impacts).
- Increases in TEV_2' and E_2' indirectly <u>change</u> M_2' ($\pm V_{ME}$).
- Changes in M_2' indirectly <u>change</u> μ_2' ($\pm V_{IME}$).

Consider an investment in HotelCo, just like AlphaCo in prior chapters, but with a business unit divestment valued at $20. Proceeds include $5 to the Fund, $5 to co-investors, and the buyer's assumption of $10 of company debt. The business unit has $10 of revenue, $6 of gross profit, and $2.5 of EBITDA. The effective value creation input variables would be as follows:

Table 13.7 HotelCo effective value creation input variables after divestment

	t_1	t_2	t_{out}	t_2'	ΔX	$\overline{X}$	χ_X
FEqV	*$44*	*$88*	*+$5*	*$93*	*$49*	-	*2.11x*
TEqV	*$80*	*$180*	*+$10*	*$190*	*$110*	*$135*	*2.38x*
FO	*55%*	*48.9%*	*Indirect*	*48.9%*	*-6.1%*	*52.0%*	*0.89x*
ND	*$20*	*$0*	*+$10*	*$10*	*-$10*	-	-
TEV	*$100*	*$180*	*Indirect*	*$200*	*$100*	*$150*	*2.00x*
ER	*80%*	*100%*	*Indirect*	*95%*	*15.0%*	*87.5%*	*1.19x*
E	*$20*	*$30*	*+$2.5*	*$32.5*	*$12.5*	*$26*	*1.63x*
M	*5.0x*	*6.0x*	*Indirect*	*6.2x*	*1.2x*	*5.6x*	*1.23x*
R	*$80*	*$100*	*+$10*	*$110*	*$30*	*$95*	*1.38x*
EM	*25%*	*30%*	*Indirect*	*30%*	*5%*	*27.3%*	*1.18x*
GP	*$48*	*$62*	*+$6*	*$68*	-	-	-
GM	*60%*	*62%*	*Indirect*	*62%*	*2%*	-	*1.07x*
OM	*35%*	*32%*	*Indirect*	*32%*	*-3%*	-	*1.11x*
MM	*7.0x*	*8.0x*	-	*8.0x*	*1.0x*	*7.5x*	*1.14x*
μ	*71.4%*	*75.0%*	*Indirect*	*76.9%*	*5.5%*	*74.2%*	*1.08x*
β_R	-	*10.0%*	-	*10.0%*	-	-	*1.10x*
α_R	-	*13.6%*	*Indirect*	*25.0%*	-	-	*1.25x*

These effective (t_2') value creation input variables provide the measures of ΔX and $\overline{X}$ needed for the Conventional and Derivative models and the measures of χ_X required for the Logarithmic Model.

13.8. Multiple Inflows and Outflows

Investments often experience more than one holding period event. In such cases, effective entry (t_1') values must include all subsequent inflows ($t_{in(j)}$), and effective exit (t_2') values must include all preceding outflows ($t_{out(k)}$), so:

$$\Delta FEqV = FEqV_2 + \sum_k FEqV_{out(k)} - FEqV_1 - \sum_j FEqV_{in(j)} = FEqV_2' - FEqV_1'$$

For example, consider IndiaCo, just like AlphaCo in prior chapters, but with a follow-on equity investment, subsequent debt investment, and add-on acquisition as described in the sections above. The effective Fund equity value at entry would be:

$$FEqV_1' = FEqV_1 + \sum_j FEqV_{in(j)}$$

$$= FEqV_1 + FEqV_{in}^{Equity\ Invest} + FEqV_{in}^{Debt\ Invest} + FEqV_{in}^{Add\text{-}on}$$

$$= \$44 + \$5 + \$0 + \$5 = \$54$$

If IndiaCo also has an equity distribution, dividend recapitalization, and divestment as described in the sections above, the effective Fund equity value at exit would be:

$$FEqV_2' = FEqV_2 + \sum_k FEqV_{out(k)}$$

$$= FEqV_2 + FEqV_{out}^{Equity\ Dist} + FEqV_{out}^{Div\ Recap} + FEqV_{out}^{Divest}$$

$$= \$88 + \$5 + \$10 + \$5 = \$108$$

The rest of IndiaCo's value creation input variables must be adjusted in the same fashion. Effective t_1' and t_2' values are determined first for the directly-adjusted input variables ($FEqV$, $TEqV$, ND, E, R, and GP), and then these are used to determine the indirectly-adjusted input variables (FO, TEV, ER, M, EM, GM, OM, μ, and α_R).

Table 13.8 IndiaCo effective value creation input variables

Event:		Equity Invest	Debt Invest	Add-on			Equity Dist	Div Recap	Divest				
Section:		13.2	13.3	13.4			13.5	13.6	13.7				
	t_1	t_{in}	t_{in}	t_{in}	t'_1	t_2	t_{out}	t_{out}	t_{out}	t'_2	ΔX	$\bar{X}$	χ_X
FEqV	$44	+$5	-	+$5	$54	$88	+$5	+$10	+$5	$108	$54	-	2.00x
TEqV	$80	+$10	-	+$10	$100	$180	+$10	+$20	+$10	$220	$120	$160	2.20x
FO	55%	Ind	-	Ind	54%	49%	Ind	Ind	Ind	49%	-5%	52%	0.91x
ND	$20	-	+$20	+$10	$50	$0	-	-$20	+$10	-$10	-$60	-	-
TEV	$100	Ind	Ind	Ind	$150	$180	Ind	-	Ind	$210	$60	$180	1.40x
ER	80%	Ind	Ind	Ind	67%	100%	Ind	Ind	Ind	105%	38%	86%	1.57x
E	$20	-	-	+$2.5	$23	$30	-	-	+$2.5	$32.5	$10	$28	1.44x
M	5.0x	Ind	Ind	Ind	6.7x	6.0x	Ind	-	Ind	6.5x	-0.2x	6.6x	0.97x
R	$80	-	-	+$10	$90	$100	-	-	+$10	$110	$20	$100	1.22x
EM	25%	-	-	Ind	25%	30%	-	-	Ind	30%	5%	27%	1.18x
GP	$48	-	-	+$6	$54	$62	-	-	+$6	$68	-	-	-
GM	60%	-	-	Ind	60%	62%	-	-	Ind	62%	2%	-	1.07x
OM	35%	-	-	Ind	35%	32%	-	-	Ind	32%	-3%	-	1.11x
MM	7.0x	-	-	-	7.0x	8.0x	-	-	-	8.0x	1.0x	7.5x	1.14x
μ	71%	Ind	Ind	Ind	95%	75%	Ind	-	Ind	81%	-14%	88%	0.85x
β_R	-	-	-	-	-	-	-	-	-	10%	-	-	1.10x
α_R	-	-	-	-	-	-	-	-	Ind	11%	-	-	1.11x

The combined impact of inflows and outflows provide $FEqV'_1$ of $54 and $FEqV'_2$ of $108 for Fund-level absolute ($) value creation of $54 and a Gross MOIC of 2.00x. IndiaCo's resulting value creation analysis is illustrated in Figure 13.2. The initial investment and three subsequent events provide three equity inflows that add up to $54. The exit and three preceding events provide four equity outflows that add up to $108. Fund value creation makes up the difference, with positive and negative value creation components "bridging the gap" between $FEqV'_1$ and $FEqV'_2$.

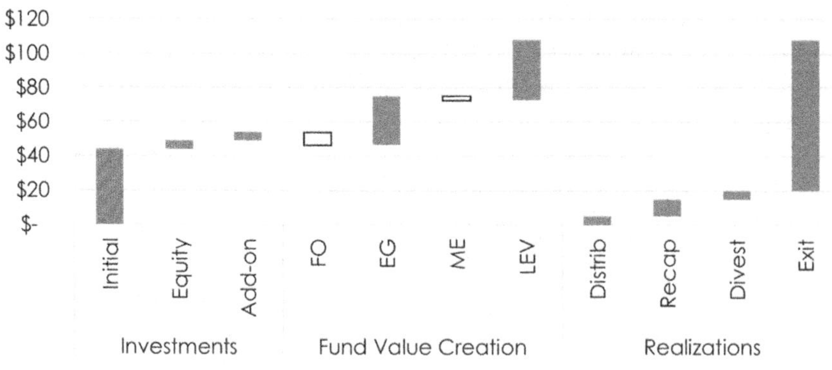

Figure 13.2 Value creation bridge for IndiaCo

The bookkeeping process behind Table 13.8 is tedious, but it is necessary to lay out every value creation input and think through how the inflow or outflow may impact each variable. Otherwise, required adjustments may be neglected. For example, because raising third-party debt involves no investment from the Fund or co-investors, it is easy to neglect that the event impacts TEV_1', ER_1', M_1', and μ_1'.

After accounting for all appropriate inflow and outflow adjustments, the effective t_1' and t_2' values provide the necessary ΔX, $\overline{X}$, and χ_X values to calculate absolute ($) value creation, relative (%) value creation, and equity return multiples or multipliers using any of the three value creation models. However, value creation IRRs are also influenced by the timing of inflows and outflows. This is addressed in the next section.

13.9. The Effective Hold

Prior chapters described value creation IRRs in terms of their equity return multipliers. For example, the Gross IRR and value creation IRR driven by EBITDA Growth would be described:

$$IRR_{GROSS} = \chi_{GROSS}{}^{1/\Delta t} - 1 = (FEqV_2'/FEqV_1')^{1/\Delta t} - 1$$

$$IRR_{EG} = \chi_{EG}{}^{1/\Delta t} - 1 = (E_2'/E_1')^{1/\Delta t} - 1$$

These expressions, however, are inaccurate for investments with more than two cash flows. Additional inflows or outflows change the Gross IRR and, therefore, every other component value creation IRR. To illustrate the impact of inflows and outflows on IRRs, consider the ten scenarios in Table 13.9. Each has a $FEqV_1'$ of $10, $FEqV_2'$ of $20, and Gross MOIC of 2.0x, but variations in cash flow timing change their Gross IRRs.

Table 13.9 Various 2.0x MOIC scenario cash flows and IRRs

	Scenario									
	A	B	C	D	E	F	G	H	I	J
12/31/2013	-$10	-$5	-$10	-$5	-$10	-$5	-$10	-$2.5	-$10	-$5
12/31/2014	-	-$5	-	-	-	-	+$10	-$2.5	+$5	+$10
12/31/2015	-	-	-	-$5	+$10	-	-	-$2.5	+$5	-
12/31/2016	-	-	+$10	-	-	-$5	-	-$2.5	+$5	-$5
12/31/2017	+$20	+$20	+$10	+$20	+$10	+$20	+$10	+$20	+$5	+$10
Invested	$10	$10	$10	$10	$10	$10	$10	$10	$10	$10
Realized	$20	$20	$20	$20	$20	$20	$20	$20	$20	$20
MOIC	2.0x	2.0x	2.0x	2.0x	2.0x	2.0x	2.0x	2.0x	2.0x	2.0x
IRR	18.9%	21.7%	22.1%	24.9%	27.2%	28.4%	38.0%	29.7%	34.9%	100%

The IRR_{GROSS} formula above provides the lowest positive IRR that is theoretically achievable. It assumes that all inflows occur at t_1 and outflows occur at t_2, which is effectively Scenario A. Every other scenario must have a higher IRR because a 2.0x MOIC is achieved with less capital "in play" or "at risk" during the four-year hold. Positive IRRs increase as capital is invested more slowly or realized more quickly. The highest IRR occurs in Scenario J, which is effectively two separate 2.0x MOIC investments, one in the first year of the hold and another in the last year of the hold. They would each have IRRs of 100%, and their combined IRR is also 100%.

Consider how the concept of capital "in play" or "at risk" might be embedded into the IRR_{GROSS} formula through an adjustment to Δt. Scenario B has an initial investment of \$5.0 and a follow-on investment of \$5.0 one year later. Effectively, half of invested capital is in play for 4.0 years (2013 through 2017) and the other half is in play for 3.0 years (2014 through 2017). This suggests a weighted average hold ($\Delta t_{WEIGHTED}$) of about 3.5 years. Similarly, Scenario C realizes \$10.0 at exit after a prior realization of \$10.0 one year earlier. Effectively, half of realized capital is in play for 4.0 years (2013 through 2017) and the other half is in play for 3.0 years (2013 through 2016). This also suggests a $\Delta t_{WEIGHTED}$ value of about 3.5 years. The $\Delta t_{WEIGHTED}$ values are dollar-weighted durations, and substituting them for Δt in the IRR formula provides:

$$IRR = MOIC^{1/\Delta t_{WEIGHTED}} - 1$$

$$= 2.0x^{1/3.5} - 1 = 21.9\%$$

This value falls within 0.2% of the IRRs for Scenario B and Scenario C in Table 13.9.

Unfortunately, errors in $\Delta t_{WEIGHTED}$-driven IRRs tend to increase as inflows occur later and outflows occur earlier in the hold. For example, both Scenario D and Scenario E have a $\Delta t_{WEIGHTED}$ value of 3.0 years, suggesting an IRR of 26.0% (±1.2% error). Both Scenario F and Scenario G have a $\Delta t_{WEIGHTED}$ value of 2.5 years, suggesting an IRR of 32.0% (±6.0% error). The bigger problem with $\Delta t_{WEIGHTED}$-driven IRRs, however, is that calculations become unwieldy and impractical when there are numerous inflows or outflows. Assuming the underlying cashflows are available for every investment of interest, which is not always the case, it may take an entire afternoon to determine the $\Delta t_{WEIGHTED}$ value for a small portfolio, and that value must be recalculated every quarter, regardless of whether there are new inflows or outflows.

A better approach uses each investment's reported Gross MOIC and Gross IRR to approximate Δt. These values are almost always available in Fund reporting and GP marketing materials. Rearranging the IRR_{GROSS} formula and taking the natural logarithm of both sides provides:

$$1 + IRR_{GROSS} = \chi_{GROSS}{}^{1/\Delta t}$$

$$\ln(1 + IRR_{GROSS}) = \ln\left(\chi_{GROSS}{}^{1/\Delta t}\right) = \frac{\ln(\chi_{GROSS})}{\Delta t}$$

The Δt value can be isolated because $\ln(x^y) = y \cdot \ln(x)$. Rearranging the formula and defining Δt as the Effective Hold ($\Delta t'$) provides a dollar-weighted duration that relies on only the Gross MOIC and Gross IRR:

$$\Delta t' = \frac{\ln(\chi_{GROSS})}{\ln(1 + IRR_{GROSS})}$$

Effective Holds for the scenarios in Table 13.9 are provided below. Every scenario has a Δt of 4.0 years. When investments have only two cash flows, like Scenario A, $\Delta t'$ converges to Δt. Whenever investments have multiple inflows and outflows, $\Delta t'$ is less than Δt because the total amount of investment capital is not "in play" or "at risk" for the entire hold.

Table 13.10 Scenario effective holds

	Scenario									
	A	B	C	D	E	F	G	H	I	J
MOIC	2.0x	2.0x	2.0x	2.0x	2.0x	2.0x	2.0x	2.0x	2.0x	2.0x
IRR	18.9%	21.7%	22.1%	24.9%	27.2%	28.4%	38.0%	29.7%	34.9%	100%
$\Delta t'$	4.00	3.53	3.48	3.11	2.88	2.78	2.15	2.66	2.32	1.00

These $\Delta t'$ values should, therefore, replace Δt in every value creation IRR formula described in *Private Equity Value Creation Analysis.* For example, the relationship between an investment's Gross IRR and the value creation IRRs for EBITDA Growth, Multiple Expansion, Leverage Effect, and Fund Ownership Impact can be described as follows:

$$(1 + IRR_{GROSS}) = (1 + IRR_{EG}) \cdot (1 + IRR_{ME}) \cdot (1 + IRR_{LEV}) \cdot (1 + IRR_{FO})$$

$$MOIC^{1/\Delta t'} = \chi_{EG}{}^{1/\Delta t'} \cdot \chi_{ME}{}^{1/\Delta t'} \cdot \chi_{LEV}{}^{1/\Delta t'} \cdot \chi_{FO}{}^{1/\Delta t'}$$

Are the $\Delta t'$-driven value creation IRRs perfect? No. The Effective Hold provides a perfectly accurate description of changes in $FEqV$ because $\Delta t'$ is derived from changes in $FEqV$. The inflows and outflows underlying IRR_{EG}, IRR_{ME}, IRR_{LEV}, and IRR_{FO} may follow a different path. Theoretically, each value creation component has its own $\Delta t'$, and the value creation IRR formulas should be expressed:

$$(1 + IRR_{GROSS}) = (1 + IRR_{EG}) \cdot (1 + IRR_{ME}) \cdot (1 + IRR_{LEV}) \cdot (1 + IRR_{FO})$$

$$MOIC^{1/\Delta t'_{FEqV}} = x_{EG}{}^{1/\Delta t'_{EG}} \cdot x_{ME}{}^{1/\Delta t'_{ME}} \cdot x_{LEV}{}^{1/\Delta t'_{LEV}} \cdot x_{FO}{}^{1/\Delta t'_{FO}}$$

For example, if the second cash flow in Scenario B is a follow-on equity investment, it should have no impact on the company P&L. From an IRR_{EG} perspective, $\Delta t'_{EG}$ is 4.0 years because the inflow has no impact on EBITDA. However, the inflow directly impacts $FEqV$ and $TEqV$, and indirectly impacts M, ER, and FO. Therefore, $\Delta t'_{ME}$, $\Delta t'_{LEV}$, and $\Delta t'_{FO}$ would be different and probably lower than the $\Delta t'_{FEqV}$ value of 3.53 years to compensate for the higher $\Delta t'_{EG}$ of 4.00 years.

However, the additional precision gained by calculating each value driver's Effective Hold is probably not worth the effort. Inflow and outflow timing have no impact on equity return multiples or multipliers, absolute ($) value creation, or relative (%) value creation. Variations in $\Delta t'$ provide only minor adjustments to results that are principally determined by equity return multipliers. Therefore, the $\Delta t'$ values calculated from the investment's Gross MOIC and Gross IRR can generally be applied to all value creation components.

13.10. Application: Effective Hold

Consider the Reported Fund track record in Table 13.11.

Table 13.11 Reported Fund returns

	Investment Dates		Capital Invested	Cash Realized	Current Value	Total Value	Gross MOIC	Gross IRR
	Entry	Exit						
JuliettCo	Dec-15	Jun-18	$15.0	$45.0	$0.0	$45.0	3.00x	64.2%
KiloCo	Jun-16	Jun-19	5.0	0.5	0.0	0.5	0.10x	-53.6%
Gross Realized Subtotal			$20.0	$45.5	$0.0	$45.5	2.28x	46.2%
	Entry	Current						
LimaCo	Dec-16	Dec-19	$15.0	$0.0	$30.0	$30.0	2.00x	26.0%
MikeCo	Jun-17	Dec-19	15.0	0.0	16.0	16.0	1.07x	3.3%
NovemCo	Dec-17	Dec-19	10.0	0.0	7.5	7.5	0.75x	-13.4%
OscarCo	Jun-18	Dec-19	15.0	0.0	25.0	25.0	1.67x	40.4%
Gross Unrealized Subtotal			$55.0	$0.0	$78.5	$78.5	1.43x	17.8%
Total Gross Returns			$75.0	$45.5	$78.5	$124.0	1.65x	28.2%
Total Net Returns							1.40x	19.0%

Just looking at the difference between entry and exit dates suggests a holding period of 2.5 years for JuliettCo and 3.0 years for KiloCo. However, JuliettCo's $\Delta t'$, as determined by $\ln(3.00x)/\ln(1 + 64.2\%)$, is 2.22 years. The lower Effective Hold reflects an inflow or outflow that is not immediately apparent in Table 13.11. Therefore, inflow or outflow adjustments to JuliettCo's value creation input variables are required. In contrast, KiloCo's $\Delta t'$, as determined by $\ln(0.10x)/\ln(1 - 53.6\%)$, matches the apparent holding period of 3.00 years. This indicates that all KiloCo cash flows occur on the entry and exit dates.

JuliettCo's value creation IRR formulas would use the $\Delta t'$ of 2.22 years. For example, if JuliettCo's EBITDA grows by 50% over the hold, providing a χ_{EG} of 1.50x, its EBITDA Growth-driven value creation IRR is:

$$IRR_{EG} = \chi_{EG}{}^{1/\Delta t'} - 1 = 1.50x^{1/2.22} - 1 = 20.0\%$$

Value creation IRRs for the other value creation components are calculated in the same manner.

Using Effective Holds saves significant time when evaluating groups of investments like the Total or Subtotals in Table 13.11. What would the entire portfolio's Δt be? The time from the first investment to 12/31/2019 is 4.0 years. With some effort, it can be shown that the average hold is 2.42 years and the weighted average hold is 2.37 years. The entire portfolio's Effective Hold can be determined with only two input variables:

$$\Delta t' = \frac{\ln(\chi_{GROSS})}{\ln(1 + IRR_{GROSS})} = \frac{\ln(1.65x)}{\ln(1 + 28.2\%)} = 2.02\ years$$

The Fund's $\Delta t'$ is about half of its Δt of 4.0 years because only about half of the Fund's Capital Invested is "in play" or "at risk" during Δt. Based on the entry and exit dates, it takes 1.5 years for the portfolio to surpass 50% invested, and the first portfolio exits begin occurring after 2.5 years. The Effective Hold captures these dynamics in a single number. Moreover, $\Delta t'$ is easily calculated from information that is almost always available in Fund reporting and GP marketing materials.

Since a portfolio's $\Delta t'$ value is a manifestation of its cash flow dynamics, it can be used to determine portfolio-level value creation IRRs. For example, if total EBITDA grows by 25% across the portfolio, the portfolio-level EBITDA Growth-driven value creation IRR is:

$$IRR_{EG} = \chi_{EG}{}^{1/\Delta t'} - 1 = 1.25x^{1/2.02} - 1 = 11.7\%$$

Portfolio-level Effective Holds will prove particularly useful in the next chapter, where the impact of Fees & Expenses and Carried Interest are modeled as value drivers in order to estimate Net Returns for private company investments and portfolios.

13.11. Summary

Holding period events do not change any of the 200+ value creation formulas, but they do change where each analysis begins and ends. Inflows like follow-on equity investments, subsequent debt investments, and add-on acquisitions change an investment's effective starting point. Outflows like equity distributions, dividend recapitalizations, and divestments change an investment's effective ending point.

Getting all the value creation input variables right requires carefully thinking through how an event impacts every input. This is especially true for investments with multiple inflows and outflows, like IndiaCo in Section 13.8. Balance sheet and P&L input variables like $FEqV$, $TEqV$, ND, E, R, and GP must be adjusted directly. Other input variables like FO, TEV, ER, M, EM, GM, OM, μ, and α_R change indirectly because they are downstream from the direct adjustments.

Return and value driver measurements like absolute ($) value creation, relative (%) value creation, and equity return multiples or multipliers do not depend on the timing of holding period events. Value creation IRRs, however, do depend on timing and should be calculated using the Effective Hold ($\Delta t'$), as described in Section 13.9.

The next chapter completes the theoretical outline of *Private Equity Value Creation Analysis* with an investigation of private company and portfolio Net Returns. By treating Fees & Expenses and Carried Interest as value drivers that "bridge the gap" between each investment's Gross Return and Net Return, an analytical framework is established that provides Net MOICs and Net IRRs, without the need of time series data or complicated cash flow models.

13.12. Optional: Holding Period Events as Value Drivers

Private Equity Value Creation Analysis does not describe events like add-on acquisitions or dividend recapitalizations as value drivers, but some analytical frameworks do. This section suggests a few comparable options that work within the framework of the Conventional, Derivative, and Logarithmic models.

When investments include add-on acquisitions, it is common to ask how much EBITDA Growth is "organic" versus "inorganic". Presumably, GPs should get more credit for growth of the organic variety. The models in this volume are fundamentally organic because adjustments for holding period events eliminate any of the inorganic components. For example, any acquired EBITDA (E_{in}) would be deducted from the value creation input variable ΔE, as follows:

$$\Delta E = E_2' - E_1' = E_2 - E_1 - E_{in}$$

However, consider a situation where after the add-on acquisition occurs the platform entity drives 50% of ΔE and the add-on entity drives 50% of ΔE. The platform entity's EBITDA growth is defined as ΔE_{PLAT}, and the add-on entity's EBITDA growth is defined as ΔE_{AO}, where:

$$\Delta E = \Delta E_{PLAT} + \Delta E_{AO}$$

Since ΔE_{AO} is a measure of growth that occurs after the add-on entity is acquired, it remains organic growth. These variables may be substituted into the Derivative Model formula for EBITDA Growth to provide:

$$\begin{aligned} V_{EG} &= \Delta E \cdot \overline{M'} \cdot \overline{ER'} \cdot \overline{FO'} \\ &= (\Delta E_{PLAT} + \Delta E_{AO}) \cdot \overline{M'} \cdot \overline{ER'} \cdot \overline{FO'} \\ &= \Delta E_{PLAT} \cdot \overline{M'} \cdot \overline{ER'} \cdot \overline{FO'} + \Delta E_{AO} \cdot \overline{M'} \cdot \overline{ER'} \cdot \overline{FO'} \\ &= V_{PLAT-EG} + V_{AO-EG} \end{aligned}$$

This expression represents absolute ($) value creation driven by platform EBITDA Growth ($V_{PLAT-EG}$) and add-on EBITDA Growth (V_{AO-EG}). Removing $\overline{ER'}$ from the expressions above provides the Conventional Model formulas. Note the superscript ($'$) which indicates that the value creation input variables for $\overline{M'}$, $\overline{ER'}$, and $\overline{FO'}$ must be fully adjusted for the add-on acquisition, as described in Section 13.4.

Using the input variables in Table 13.4 to solve for EchoCo provides:

$$\begin{aligned} V_{EG} &= V_{PLAT-EG} + V_{AO-EG} \\ &= (50\% \cdot \$7.5) \cdot 5.67x \cdot 87.5\% \cdot 51.7\% + (50\% \cdot \$7.5) \cdot 5.67x \cdot 87.5\% \cdot 51.7\% \\ &= \$9.6 + \$9.6 = \$19.2 \end{aligned}$$

Because V_{EG} is directly proportional to ΔE, in the case of the Conventional and Derivative models, any factor that contributes 50% to ΔE is also responsible for 50% of V_{EG}.

A solution for the Logarithmic Model requires the approach taken in Section 9.3 for Gross Margin Expansion and Operating Margin Reduction. Note that the continuous change in EBITDA can be estimated by:

$$r_{EG} = \ln(E_2'/E_1') \approx \frac{E_2' - E_1'}{(E_1' + E_2')/2} = \frac{\Delta E}{\overline{E'}}$$

$$= \frac{\Delta E_{PLAT} + \Delta E_{AO}}{\overline{E'}} = \frac{\Delta E_{PLAT}}{\overline{E'}} + \frac{\Delta E_{AO}}{\overline{E'}}$$

EBITDA Growth's absolute ($) value creation given by the Logarithmic Model can, therefore, be expressed:

$$V_{EG} = \Delta FEqV \cdot \frac{\ln(E_2'/E_1')}{\ln(MOIC)} \approx \Delta FEqV \cdot \frac{\Delta E/\overline{E'}}{\ln(MOIC)}$$

$$= \Delta FEqV \cdot \frac{\Delta E_{PLAT}/\overline{E'}}{\ln(MOIC)} + \Delta FEqV \cdot \frac{\Delta E_{AO}/\overline{E'}}{\ln(MOIC)}$$

$$= V_{PLAT-EG} + V_{AO-EG}$$

Using the input variables in Table 13.4 to solve for EchoCo provides:

$$V_{EG} = V_{PLAT-EG} + V_{AO-EG}$$

$$= \$39 \cdot \frac{(50\% \cdot \$7.5)/\$26.25}{\ln(1.80x)} + \$39 \cdot \frac{(50\% \cdot \$7.5)/\$26.25}{\ln(1.80x)}$$

$$= \$9.5 + \$9.5 = \$19.0$$

Similarly, EBITDA Growth's equity return multipliers given by the Logarithmic Model can be expressed:

$$X_{EG} = X_{PLAT-EG} \cdot X_{AO-EG}$$

$$= X_{EG}^{\Delta E_{PLAT}/\Delta E} \cdot X_{EG}^{\Delta E_{AO}/\Delta E} \approx e^{\Delta E_{PLAT}/\overline{E'}} \cdot e^{\Delta E_{AO}/\overline{E'}}$$

$$= 1.33x^{\$3.25/\$7.50} \cdot 1.33x^{\$3.25/\$7.50} \approx e^{\$3.25/\$26.25} \cdot e^{\$3.25/\$26.25}$$

$$= 1.155x \cdot 1.155x \approx 1.154x \cdot 1.154x$$

M&A is often discussed in terms of its impact on valuation. If a t_1 entry multiple is 5.0x and a t_2 exit multiple is 6.0x, value creation driven by Multiple Expansion is proportional to ΔM of 1.0x. However, if an add-on acquisition decreases the effective entry multiple to 4.8x, ΔM would increase to 1.2x. In this case, Multiple Expansion can be modeled as +1.0x driven by the platform (ΔM_{PLAT}) and +0.2x driven by the add-on (ΔM_{AO}).

The ΔM formula may be generalized as follows:

$$\Delta M = \Delta M_{PLAT} + \Delta M_{AO}$$
$$= (M_2 - M_1) + (M_1 - M_1')$$
$$= \left(\frac{TEV_2}{E_2} - \frac{TEV_1}{E_1}\right) + \left(\frac{TEV_1}{E_1} - \frac{TEV_1 + TEV_{in}}{E_1 + E_{in}}\right)$$

Above, ΔM_{AO} is a measure of arbitrage in the entry Multiple driven by the cost of E_{in} acquired during the hold. These variables may be substituted into the Derivative Model formula for Multiple Expansion to provide:

$$V_{ME} = \Delta M \cdot \overline{E'} \cdot \overline{ER'} \cdot \overline{FO'}$$
$$= (\Delta M_{PLAT} + \Delta M_{AO}) \cdot \overline{E'} \cdot \overline{ER'} \cdot \overline{FO'}$$
$$= \Delta M_{PLAT} \cdot \overline{E'} \cdot \overline{ER'} \cdot \overline{FO'} + \Delta M_{AO} \cdot \overline{E'} \cdot \overline{ER'} \cdot \overline{FO'}$$
$$= V_{PLAT-ME} + V_{AO-ME}$$

This expression represents absolute ($) value creation driven by platform Multiple Expansion ($V_{PLAT-ME}$) and add-on Multiple Expansion (V_{AO-ME}). Removing $\overline{ER'}$ from the expressions above provides the Conventional Model formulas. The superscript (′) indicates that the input variables for $\overline{E'}$, $\overline{ER'}$, and $\overline{FO'}$ must be fully adjusted for the add-on acquisition, as described in Section 13.4.

Using the input variables in Table 13.4 to solve for EchoCo provides:

$$V_{ME} = V_{PLAT-ME} + V_{AO-ME}$$
$$= +1.00x \cdot \$26.25 \cdot 87.5\% \cdot 51.7\% - 0.33x \cdot \$26.25 \cdot 87.5\% \cdot 51.7\%$$
$$= \$11.9 - \$3.9 = \$8.0$$

Because EchoCo's add-on acquisition is valued at 8.0x the E_{in} of $2.5, it increases the effective entry Multiple from 5.00x to 5.33x, which decreases the effective ΔM from 1.00x to 0.67x and, therefore, reduces V_{ME}.

A solution for the Logarithmic Model notes that the continuous change in Multiple can be estimated by:

$$r_{ME} = \ln(M_2'/M_1') \approx \frac{M_2' - M_1'}{(M_1' + M_2')/2} = \frac{\Delta M}{\overline{M'}}$$
$$= \frac{\Delta M_{PLAT} + \Delta M_{AO}}{\overline{M'}} = \frac{\Delta M_{PLAT}}{\overline{M'}} + \frac{\Delta M_{AO}}{\overline{M'}}$$

Multiple Expansion's absolute ($) value creation given by the Logarithmic Model can, therefore, be expressed:

$$V_{ME} = \Delta FEqV \cdot \frac{\ln(M_2'/M_1')}{\ln(MOIC)} \approx \Delta FEqV \cdot \frac{\Delta M/\overline{M'}}{\ln(MOIC)}$$

$$= \Delta FEqV \cdot \frac{\Delta M_{PLAT}/\overline{M'}}{\ln(MOIC)} + \Delta FEqV \cdot \frac{\Delta M_{AO}/\overline{M'}}{\ln(MOIC)}$$

$$= V_{PLAT-ME} + V_{AO-ME}$$

Using the input variables in Table 13.4 to solve for EchoCo provides:

$$V_{ME} = V_{PLAT-ME} + V_{AO-ME}$$

$$= \$39 \cdot \frac{1.00x/5.67x}{\ln(1.80x)} + \$39 \cdot \frac{-0.33x/5.67x}{\ln(1.80x)}$$

$$= \$11.7 - \$3.9 = \$7.8$$

Similarly, Multiple Expansion's equity return multipliers given by the Logarithmic Model can be expressed:

$$x_{ME} = x_{PLAT-ME} \cdot x_{AO-ME}$$

$$= x_{ME}{}^{\Delta M_{PLAT}/\Delta M} \cdot x_{ME}{}^{\Delta M_{AO}/\Delta M} \approx e^{\Delta M_{PLAT}/\overline{M'}} \cdot e^{\Delta M_{AO}/\overline{M'}}$$

$$= 1.13x^{1.00x/0.67x} \cdot 1.13x^{-0.33x/0.67x} \approx e^{1.00x/5.67x} \cdot e^{-0.33x/5.67x}$$

$$= 1.193x \cdot 0.943x \approx 1.193x \cdot 0.943x$$

The overall impact of EchoCo's add-on acquisition, as described by this framework, is positive. The add-on decreases value creation from Multiple Expansion because it raises EchoCo's effective entry Multiple, but the impact on EchoCo's P&L more than makes up for it with the increase in value creation from EBITDA Growth. According to the Derivative Model, the net effect of the add-on acquisition on Fund-level value creation ($\Delta FEqV_{AO}$) is:

$$\Delta FEqV_{AO} = V_{AO-EG} + V_{AO-ME}$$

$$= +\$9.6 - \$3.9 = +\$5.7$$

Expressions of this form can be extended to the other value creation components. Such analyses are more likely to prove useful in specialized circumstances, like a case study, rather than a systematic framework for evaluating many companies.

The example above remains a measurement of organic growth. It does not describe, for example, how much more EchoCo is worth at exit because the add-on acquisition provided synergies, increased scale, or helped the company to offer a more attractive mix of products or services. *Private Equity Value Creation Analysis* captures such effects through the value creation component of Intrinsic Multiple Expansion, as described in Chapter 10.

14. Gross Returns and Net Returns

This chapter introduces an analytical framework that reliably estimates Net Returns for private companies and portfolios. Net IRRs are calculated using a streamlined model that requires no time series data or complicated cash flow models.

14.1. Background

While Gross Returns are the fundamental measurement of GP investment activities, Net Returns are the most important numbers to LPs. In theory, a Fund's Net Return is its portfolio-level Gross Return less Management Fees and Fund Expenses (*F&E*) and Carried Interest (*CI*). In practice, the relationship between the Net Return and Gross Return is obscured by various Fund-level dynamics, including recycling, short term portfolio investments, transaction and monitoring fee offsets, the use of Fund credit facilities, assets and liabilities on Fund balance sheets, different LPs with different terms and capital call requirements, etc.

Evaluating the Net Return for a single portfolio company or group of companies, within or across Funds, usually requires first measuring *F&E*, *CI*, and other Fund-level items, and then adjusting each company's Gross cash flow by the appropriate amounts at the appropriate times. For large portfolios, such analyses become tedious. Assumptions about allocation and timing of adjustments are subjective, and the Net Returns, especially Net IRRs, are highly sensitive to these decisions. Furthermore, the adjustments continually change as investment valuations change, so models often need to be updated, or rebuilt from scratch, on a quarterly basis.

Private Equity Value Creation Analysis takes a different approach and estimates Net Returns from the bottom-up. The *F&E* and *CI* values are viewed as value drivers that "bridge the gap" between Gross Returns and Net Returns. As in prior chapters, the value creation framework relies on accessible input variables that can usually be obtained or estimated from Fund reporting and GP marketing materials.

The notation for value creation input variables has been modified in this chapter to streamline various derivations. Fund invested capital (previously $FEqV_1'$) is defined as Gross Invested Capital (*GIC*). This equals the Net Invested Capital (*NIC*) less an investment's pro-rata share of Fund *F&E*, providing:

$$GIC = NIC - F\&E$$

Fund realized capital and unrealized value (previously $FEqV_2'$) is defined as Gross Return (*GR*). This equals the sum of Net Return (*NR*) and an investment's pro-rata share of Fund *CI*, providing:

$$GR = NR + CI$$

Gross MOIC ($MOIC_{GROSS}$ or χ_{GROSS}) is equal to *GR* divided by *GIC*, and Net MOIC ($MOIC_{NET}$ or χ_{NET}) is equal to *NR* divided by *NIC*, providing:

$$MOIC_{GROSS} = \frac{GR}{GIC} = \frac{NR + CI}{NIC - F\&E}$$

$$MOIC_{NET} = \frac{NR}{NIC} = \frac{GR - CI}{GIC + F\&E}$$

14.2. Fees & Expenses

The first step of the value creation framework is to estimate *F&E* in terms of *GIC*. The "two" in two-and-twenty is not based on *GIC*, but Fund size or Committed Capital (*CC*). It is an annual fee equal to 2% *CC* during the "investment period", which is usually the Fund's first four or five years. Later, it is reduced to around 1% *CC* for the remainder of the Fund's life. A Management Fee of 2% for the first five years and 1% for the next five years would total 15% *CC*.

Funds also have expenses related to initial setup costs and ongoing operations. Permissible expenses are specified in Fund legal agreements and generally include certain costs related to outside counsel, accountants, auditors, and other service providers, as well as reporting to LPs. They fluctuate annually and vary by the size and type of Fund, as well as each GP's approach to managing its Funds. Annual expenses are often in the range of 0.25% *CC* to 1.0% *CC*. If Fund Expenses add up to 5% *CC*, then total *F&E*

would be 20% *CC*. In theory, this leaves 80% *CC* available for portfolio investments, so *GIC* would equal 80% *CC*, and *F&E* can be defined as:

$$F\&E = 20\%\ CC = 20\%\ CC \cdot \frac{GIC}{80\%\ CC} = 25\%\ GIC$$

However, this would not account for recycling. Funds can usually reinvest capital that is realized within 12 to 24 months of an investment. Often, they can also recycle longer-term capital gains up to the amount of *F&E*. This may lead to *GIC* values larger than Funds sizes. If recycling enables a Fund to invest 115% *CC*, then *F&E* would be:

$$F\&E = 20\%\ CC \cdot \frac{GIC}{115\%\ CC} = 17.4\%\ GIC$$

Another wrinkle is Management Fee Offsets. Private company ownership can generate fees, specifically transaction fees from acquisitions, exits, or financings and monitoring fees whenever companies are performing well enough to pay them. Such fees reduce company equity value creation and, therefore, Fund Gross Returns. Because most of these fees go directly to the GP, LPs usually require that a portion of them offset Management Fees. The ability and willingness of GPs to collect fees and thus generate Management Fee Offsets can change with time and circumstance, but reductions on the order of 5% *CC* or 10% *CC* are not uncommon.

There are other complications, such as the fact the GP is also a Fund LP. GPs own 2% to 5% of their Funds and are charged their pro-rata share of Fund Expenses, but not Management Fees. However, it is unnecessary to factor such detail into analyses. The goal is to generate robust and reasonable results, including Net IRRs, without building complicated models with countless assumptions. Surveying a few of their own Funds or portfolios should give GPs and LPs a good sense of what reasonable *F&E* values should be.

Factoring in recycling and some degree of Management Fee Offsets, a good starting point might be 12% *GIC*, which could be applied by:

- Adding 12% *GIC* to every portfolio investment, regardless of how long each investment is held
- Scaling the 12% *GIC* to each investment's hold (Δt) in such a way that the combined *F&E* for all investments sums to 12% *GIC*

The first approach is simpler, but it discriminates against newer investments which are burdened with a decade of *F&E* before they have an opportunity to create value. The second method might divide 12% *GIC* by the average portfolio hold, perhaps 6.0 years, providing an annual fee of 2% *GIC*. Total *F&E* would continue to be 12% *GIC*, but investments held for two or eight years would have *F&E* of 4% *GIC* or 16% *GIC*, respectively.

14.3. Carried Interest

With reasonable *F&E* estimates in hand, the "twenty" in two-and-twenty can be determined. Carried Interest is 20% of Net Profits (*NP*), which is a simple concept but tricky to implement because *NP* is a conditional value:

$$NP = \begin{cases} GR - GIC - F\&E, & when\ GR - GIC - F\&E > \$0 \\ \$0, & when\ GR - GIC - F\&E \leq \$0 \end{cases}$$

The term $GR - GIC - F\&E$ can be negative, but from a Carried Interest perspective, the minimum value is \$0. Factoring conditionality into *CI* provides:

$$CI = 20\% \cdot NP$$

$$CI = 20\% \cdot \begin{cases} GR - GIC - F\&E, & when\ GR - GIC - F\&E > \$0 \\ \$0, & when\ GR - GIC - F\&E \leq \$0 \end{cases}$$

Conditionality can be implemented with logical functions like IF or MAX. For example, the following function returns \$0 whenever the *NP* term is negative:

$$CI = 20\% \cdot \text{MAX}[\$0, GR - GIC - F\&E]$$

Real world Carried Interest includes additional complications. Certain hurdle rates must be achieved before *CI* can be distributed to GPs, and these are not factored into the following analyses. Furthermore, investments within a Fund are cross-collateralized. *CI* depends on the entire portfolio achieving positive *NP*. Most portfolios have a handful of very good investments that generate *CI* and one or two unfortunate deals that destroy it. It typically makes sense to exclude cross-collateralization when evaluating Net Returns for individual investments, but the value creation formulas can be adjusted to either include or exclude it and provide individual investment Net Returns with either pooled or deal-by-deal carried interest models.

14.4. Conventional and Derivative Models

Gross absolute (\$) value creation (V_{GROSS}) for a Fund's investment is $\Delta FEqV$, or in this chapter's notation:

$$V_{GROSS} = GR - GIC$$

Net absolute ($) value creation ($V_{NET}$) accounts for each investment's share of Fund $F\&E$ and CI, as follows:

$$V_{NET} = NR - NIC = GR - CI - GIC - F\&E$$

In the expression above, $F\&E$ can be viewed as a Fees & Expenses value driver (V_{FEES}) and CI can be viewed as a Carried Interest value driver (V_{CARRY}), providing:

$$\begin{aligned} V_{GROSS} &= GR - GIC \\ &= (NR + CI) - (NIC - F\&E) \\ &= (NR - NIC) + F\&E + CI \\ &= V_{NET} + V_{FEES} + V_{CARRY} \end{aligned}$$

The value creation framework expresses $F\&E$ in terms of percent GIC ($\%F$) and CI in terms of percent NP ($\%C$, usually 20%), as follows:

$$V_{FEES} = \%F \cdot GIC$$

$$V_{CARRY} = \%C \cdot \mathrm{MAX}[\$0, GR - GIC \cdot (1 + \%F)]$$

Using these expressions to solve for V_{NET} provides:

$$\begin{aligned} V_{NET} &= V_{GROSS} - V_{FEES} - V_{CARRY} \\ &= GR - GIC - \%F \cdot GIC - \%C \cdot \mathrm{MAX}[\$0, GR - GIC \cdot (1 + \%F)] \end{aligned}$$

Relative (%) value creation can be determined from absolute ($) value creation:

$$v_{NET} + v_{FEES} + v_{CARRY} = 100\%$$

$$\frac{V_{NET}}{V_{GROSS}} + \frac{V_{FEES}}{V_{GROSS}} + \frac{V_{CARRY}}{V_{GROSS}} = 100\%$$

Equity return multipliers can be determined from relative (%) value creation:

$$\begin{aligned} x_{GROSS} &= x_{NET} \cdot x_{FEES} \cdot x_{CARRY} \\ &= MOIC_{GROSS}{}^{v_{NET}} \cdot MOIC_{GROSS}{}^{v_{FEES}} \cdot MOIC_{GROSS}{}^{v_{CARRY}} \end{aligned}$$

Value creation IRRs can be determined from equity return multipliers:

$$(1 + IRR_{GROSS}) = (1 + IRR_{NET}) \cdot (1 + IRR_{FEES}) \cdot (1 + IRR_{CARRY})$$

$$x_{GROSS}{}^{1/\Delta t} = x_{NET}{}^{1/\Delta t} \cdot x_{FEES}{}^{1/\Delta t} \cdot x_{CARRY}{}^{1/\Delta t}$$

Thus, this approach provides a complete description of the relationships among Gross Returns, Net Returns, Fees & Expenses, and Carried Interest. It requires only four input variables: *GR*, *GIC*, *%F*, and *%C*. Time series data and complicated cash flow models are not needed. However, this model requires multiple steps to arrive at Net MOIC or Net IRR. The Logarithmic Model can achieve results in fewer steps.

14.5. Logarithmic Model

As with the value creation models in prior chapters, the Logarithmic Model seeks to first identify value creation components in terms of their equity return multipliers. It was shown that:

$$\chi_{NET} = \frac{NR}{NIC} = \frac{GR - CI}{GIC + F\&E}$$

Substituting $\%F \cdot GIC$ for $F\&E$ provides:

$$\chi_{NET} = \frac{GR - CI}{GIC + \%F \cdot GIC} = \frac{GR - CI}{GIC \cdot (1 + \%F)}$$

If an investment has no Net Profit, CI is \$0 and the formula is reduced to:

$$\chi_{NET} = \frac{GR}{GIC} \cdot \frac{1}{1 + \%F} = \chi_{GROSS} \cdot \frac{1}{\chi_{FEES}}$$

The first term is simply the investment's Gross MOIC. The second term is a factor that defines a relationship between χ_{GROSS} and χ_{NET} after Fees & Expenses, but before Carried Interest. The value $1 + \%F$ is, in fact, the equity return multiplier for Fees & Expenses, χ_{FEES}:

$$\chi_{FEES} = 1 + \%F$$

The term χ_{GROSS}/χ_{FEES} is the MOIC after $F\&E$ but before CI. If it is less than or equal to 1.0x, then CI is \$0. If the term is greater than 1.0x, CI is greater than \$0, and equal to $\%C \cdot [GR - GIC \cdot (1 + \%F)]$. Substituting this expression into χ_{NET} provides:

$$\chi_{NET} = \frac{GR - CI}{GIC + F\&E} = \frac{GR - \%C \cdot [GR - GIC \cdot (1 + \%F)]}{GIC \cdot (1 + \%F)}$$

$$= \frac{GR - GR \cdot \%C + \%C \cdot GIC \cdot (1 + \%F)}{GIC \cdot (1 + \%F)}$$

Multiplying the numerator's third term by 1.0x in the form of *GR/GR* provides:

$$\chi_{NET} = \frac{GR}{GIC} \cdot \frac{1}{1+\%F} \cdot \left(1 - \%C + \%C \cdot \frac{1+\%F}{GR/GIC}\right)$$

$$= \chi_{GROSS} \cdot \frac{1}{\chi_{FEES}} \cdot \left(1 - \%C + \frac{\%C \cdot \chi_{FEES}}{\chi_{GROSS}}\right)$$

$$= \chi_{GROSS} \cdot \frac{1}{\chi_{FEES}} \cdot \frac{1}{\chi_{CARRY}}$$

Consider how the expression above functions. Net MOICs must be lower than Gross MOICs, so the factors $1/\chi_{FEES}$ and $1/\chi_{CARRY}$ must each be less than or equal to 1.0x and work to reduce χ_{GROSS} to χ_{NET}. The reciprocal of χ_{FEES} satisfies this condition because $\%F$ is positive, so $1/(1 + \%F)$ must always be less than or equal to 1.0x. The third term will also be less than 1.0x whenever $\chi_{GROSS} > \chi_{FEES}$ because in such cases χ_{FEES}/χ_{GROSS} is less than 1.0x. To illustrate, consider an investment with a %C of 20%, a %F of 12%, and a Gross MOIC of 2.0x:

$$\frac{1}{\chi_{CARRY}} = \left(1 - 20\% + \frac{20\% \cdot 1.12x}{2.00x}\right) = 80\% + \frac{20\%}{1.79x} = 0.91x$$

To generalize χ_{CARRY}, an operator like MAX must be included to provide conditionality and ensure that the formula returns a value greater than or equal to 1.0x, as follows:

$$\chi_{CARRY} = \text{MAX}\left[1.0x, \frac{1}{1 - \%C + \dfrac{\%C \cdot \chi_{FEES}}{\chi_{GROSS}}}\right]$$

Solving χ_{CARRY} for the investment above provides:

$$\chi_{CARRY} = \text{MAX}\left[1.0x, \frac{1}{1 - 20\% + \dfrac{20\% \cdot 1.12\text{x}}{2.00x}}\right] = 1.10x$$

The relationships among Gross Return, Net Return, Fees & Expenses, and Carried Interest for this investment, in the form of equity return multipliers, are given by:

$$\chi_{GROSS} = \chi_{NET} \cdot \chi_{FEES} \cdot \chi_{CARRY}$$
$$2.00x = 1.63x \cdot 1.12x \cdot 1.10x$$

Value creation IRRs can be determined from equity return multipliers. Assuming a three-year hold:

$$(1 + IRR_{GROSS}) = (1 + IRR_{NET}) \cdot (1 + IRR_{FEES}) \cdot (1 + IRR_{CARRY})$$

$$\chi_{GROSS}{}^{1/\Delta t} = \chi_{NET}{}^{1/\Delta t} \cdot \chi_{FEES}{}^{1/\Delta t} \cdot \chi_{CARRY}{}^{1/\Delta t}$$

$$2.0x^{1/3.0} = 1.63x^{1/3.0} \cdot 1.12x^{1/3.0} \cdot 1.10x^{1/3.0}$$

$$(1.260) = (1.177) \cdot (1.038) \cdot (1.031)$$

$$(1 + 26.0\%) = (1 + 17.7\%) \cdot (1 + 3.8\%) \cdot (1 + 3.1\%)$$

Relative (%) value creation can be determined from equity return multipliers:

$$v_{NET} + v_{FEES} + v_{CARRY} = 100\%$$

$$\frac{\ln(\chi_{NET})}{\ln(\chi_{GROSS})} + \frac{\ln(\chi_{FEES})}{\ln(\chi_{GROSS})} + \frac{\ln(\chi_{CARRY})}{\ln(\chi_{GROSS})} = 100\%$$

$$\frac{\ln(1.177)}{\ln(1.260)} + \frac{\ln(1.038)}{\ln(1.260)} + \frac{\ln(1.031)}{\ln(1.260)} = 100\%$$

$$70.4\% + 16.3\% + 13.3\% = 100\%$$

Assuming this investment generates Fund equity value growth of \$50, absolute (\$) value creation can be determined from relative (%) value creation:

$$V_{GROSS} = V_{NET} + V_{FEES} + V_{CARRY}$$

$$= v_{NET} \cdot V_{GROSS} + v_{FEES} \cdot V_{GROSS} + v_{CARRY} \cdot V_{GROSS}$$

$$= 70.4\% \cdot \$50 + 16.3\% \cdot \$50 + 13.3\% \cdot \$50$$

$$= \$35.2 + \$8.2 + \$6.6 = \$50.0$$

Thus, this approach provides a complete description of the relationships among Gross Returns, Net Returns, Fees & Expenses, and Carried Interest. It requires only three input variables: Gross MOIC, $\%F$, and $\%C$. Time series data and complicated cash flow models are not needed. Examples are provided in the following sections.

14.6. Application: Portfolio Net Returns

The track record in Table 14.1 is typical of the format found in most Fund reporting and GP marketing materials. It provides Net Returns for only the entire Fund, but Gross Returns for individual investments like JuliettCo and KiloCo, as well as groups of investments like the Realized Subtotal and Unrealized Subtotal.

Table 14.1 Reported Fund returns

	Investment Dates		Capital Invested	Cash Realized	Current Value	Total Value	Gross MOIC	Gross IRR
	Entry	Exit						
JuliettCo	Dec-15	Jun-18	$15.0	$45.0	$0.0	$45.0	3.00x	64.2%
KiloCo	Jun-16	Jun-19	5.0	0.5	0.0	0.5	0.10x	-53.6%
Gross Realized Subtotal			$20.0	$45.5	$0.0	$45.5	2.28x	46.2%
	Entry	Current						
LimaCo	Dec-16	Dec-19	$15.0	$0.0	$30.0	$30.0	2.00x	26.0%
MikeCo	Jun-17	Dec-19	15.0	0.0	16.0	16.0	1.07x	3.3%
NovemCo	Dec-17	Dec-19	10.0	0.0	7.5	7.5	0.75x	-13.4%
OscarCo	Jun-18	Dec-19	15.0	0.0	25.0	25.0	1.67x	40.4%
Gross Unrealized Subtotal			$55.0	$0.0	$78.5	$78.5	1.43x	17.8%
Total Gross Returns			$75.0	$45.5	$78.5	$124.0	1.65x	28.2%
Total Net Returns							1.40x	19.0%

A reasonable Fees & Expenses estimate for the Fund will be sufficient to provide a Net Return for each item in the table above. If the Reported Fund's *F&E* was 3% *GIC* in the first year and 2% *GIC* in each of the following 3.5 years, then *F&E* would total 10% *GIC*, or $7.5. Assuming a %*C* of 20%, the Gross MOICs would provide the following χ_{CARRY} and Net MOIC values:

Table 14.2 Calculation of Net MOICs for the Reported Fund

	Gross MOIC	χ_{FEES}	χ_{CARRY}	Net MOIC
JuliettCo	3.00x	1.10x	1.15x	2.38x
KiloCo	0.10x	1.10x	1.00x	0.09x
Realized Subtotal	2.28x	1.10x	1.12x	1.85x
LimaCo	2.00x	1.10x	1.10x	1.65x
MikeCo	1.07x	1.10x	1.00x	0.97x
NovemCo	0.75x	1.10x	1.00x	0.68x
OscarCo	1.67x	1.10x	1.07x	1.41x
Unrealized Subtotal	1.43x	1.10x	1.05x	1.24x
Fund Total	1.65x	1.10x	1.07x	1.40x

In Table 14.2, each χ_{CARRY} and Net MOIC value is calculated from the Gross MOIC and χ_{FEES} value to its immediate left. Therefore, Net MOICs for individual investments are deal-by-deal returns (non-collateralized), Subtotals are based on the combined net profits of investments in each group (partially collateralized), and the Fund Total is based on net profits of all six investments (cross-collateralized). Investments without Carried Interest have the minimum χ_{CARRY} value of 1.0x and Net MOICs less than or equal to 1.0x. This happens whenever Gross MOIC is less than or equal to χ_{FEES}. For example, even though MikeCo's Gross MOIC is greater than 1.0x, it has no Carried Interest because its appreciation of +$1.0 is less than its *F&E* of $1.5.

The Reported Fund's underlying Gross and Net cash flows are provided below for comparison purposes.

Table 14.3 Underlying Gross cash flows of the Reported Fund

	JuliettCo	KiloCo	LimaCo	MikeCo	NovemCo	OscarCo	Total
Dec-15	-10.0						-10.0
Jun-16		-5.0					-5.0
Dec-16	-5.0		-15.0				-20.0
Jun-17				-10.0			-10.0
Dec-17					-10.0		-10.0
Jun-18	45.0					-15.0	30.0
Dec-18				-5.0			-5.0
Jun-19		0.5					0.5
Dec-19			30.0	16.0	7.5	25.0	78.5
IRR	64.2%	-53.6%	26.0%	3.3%	-13.4%	40.4%	28.2%

With Fees & Expenses of $7.5, Carried Interest for the Reported Fund would be 20% of net profits, or $8.3. If *F&E* is deducted from the Gross cash flows semi-annually while *CI* is deducted from the 12/31/2019 market value, the Reported Fund's Net IRR would be 19.0%.

Table 14.4 Underlying Net cash flows of the Reported Fund

	Gross Total	*F&E*	*CI*	Net Total
Dec-15	-10.00	-1.13		-11.13
Jun-16	-5.00	-1.13		-6.13
Dec-16	-20.00	-0.75		-20.75
Jun-17	-10.00	-0.75		-10.75
Dec-17	-10.00	-0.75		-10.75
Jun-18	30.00	-0.75		29.25
Dec-18	-5.00	-0.75		-5.75
Jun-19	0.50	-0.75		-0.25
Dec-19	78.50	-0.75	-8.30	69.45
IRR	28.2%			19.0%

Returning to the value creation framework, Net IRRs are estimated using the Table 14.2 Net MOICs and each investment's hold, Δt, in years:

$$IRR_{NET} = MOIC_{NET}{}^{1/\Delta t} - 1$$

However, as described in Section 13.9, this formula is inaccurate for any investment with more than two cash flows. The Δt values would provide Net IRRs that are too low for JuliettCo and not low enough for MikeCo. It is also unclear what the appropriate Δt would be for groups of investments, like the Realized Subtotal, Unrealized Subtotal, and Fund Total. Average hold? Weighted average hold?

A solution comes from the analysis of holding period events in Chapter 13. While a traditional hold (Δt) is defined by entry and exit dates, an Effective Hold ($\Delta t'$) can be determined from each investment's Gross MOIC and Gross IRR, as follows:

$$\Delta t' = \frac{\ln(MOIC_{GROSS})}{\ln(1 + IRR_{GROSS})}$$

The Effective Hold is a dollar-weighted duration. When investments have only two cash flows, like KiloCo, LimaCo, NovemCo, and OscarCo, $\Delta t'$ converges to Δt. Whenever investments have multiple inflows or outflows, $\Delta t'$ is less than Δt because the full amount of capital has not been "in play" or "at risk" for the entire hold.

To illustrate, consider MikeCo's two investment inflows. The initial investment of \$10 is held for 2.5 years, while the follow-on investment of \$5 is held for 1.0 years. The weighted average investment hold is therefore:

$$\Delta t_{WEIGHTED} = \frac{\$10}{\$15} \cdot 2.5\ years + \frac{\$5}{\$15} \cdot 1.0\ years = 2.00\ years$$

Compare this to the Effective Hold formula, which provides:

$$\Delta t' = \frac{\ln(MOIC_{GROSS})}{\ln(1 + IRR_{GROSS})} = \frac{\ln(1.067x)}{\ln(1 + 3.26\%)} = 2.01\ years$$

Using only the Gross MOIC and Gross IRR, which are almost always provided in Fund reporting and GP marketing materials, $\Delta t'$ gets to the right answer with no information about the date of the follow-on investment or how much of the \$15 is invested in each round. Furthermore, the $\Delta t'$ formula also works for Subtotals and Totals, where $\Delta t_{WEIGHTED}$ calculations would be impractical, if not impossible.

The Effective Holds in Table 14.5 are determined from each investment's Gross MOIC and Gross IRR in Table 14.1 and are used to calculate each investment's Net IRR as follows:

$$IRR_{NET} = MOIC_{NET}{}^{1/\Delta t'} - 1 = MOIC_{NET}{}^{\ln(1+IRR_{GROSS})/\ln(MOIC_{GROSS})} - 1$$

Table 14.5 Calculation of Net IRRs

	Investment Dates		Hold	Effective Hold	Net MOIC	Net IRR
	Entry	Exit				
JuliettCo	Dec-15	Jun-18	2.50	2.22	2.38x	47.9%
KiloCo	Jun-16	Jun-19	3.00	3.00	0.09x	-55.0%
Realized Subtotal			n/a	2.16	1.85x	33.0%
	Entry	Current				
LimaCo	Dec-16	Dec-19	3.00	3.00	1.65x	18.3%
MikeCo	Jun-17	Dec-19	2.50	2.01	0.97x	-1.5%
NovemCo	Dec-17	Dec-19	2.00	2.00	0.68x	-17.4%
OscarCo	Jun-18	Dec-19	1.50	1.50	1.41x	25.8%
Unrealized Subtotal			n/a	2.18	1.24x	10.3%
Total Net Returns			n/a	2.02	1.40x	18.2%

The value creation framework's Net IRR of 18.2% compares well to the cash flow-driven Net IRR of 19.0% provided in Table 14.4. This is not bad considering the formulas require only four input variables: Gross MOIC, Gross IRR, $\%C$, and $\%F$. While cash flow models can require complicated spreadsheets with hundreds of entries, the value creation framework can be performed in a single step, as follows:

$$IRR_{NET} = MOIC_{NET}{}^{1/\Delta t'} - 1 = \left\{\frac{X_{GROSS}}{X_{FEES} \cdot X_{CARRY}}\right\}^{1/\Delta t'} - 1$$

$$= \left\{\frac{MOIC_{GROSS}}{(1+\%F)} \cdot MIN\left[1{,}1 - \%C + \frac{\%C \cdot (1+\%F)}{MOIC_{GROSS}}\right]\right\}^{\frac{\ln(1+IRR_{GROSS})}{\ln(MOIC_{GROSS})}} - 1$$

This is, admittedly, the ugliest equation in *Volume I*, but since it can be condensed into a single spreadsheet formula, it can be easily duplicated and scaled to analyze the Net IRRs for dozens or hundreds of private company investments and portfolios.

14.7. Application: Synthetic Funds

The Net Return model of *Private Equity Value Creation Analysis* is particularly useful for evaluating new GPs or those with material team or strategy changes. New GPs market track records of investments made by professionals at one or more previous firms. GPs in transition market only portions of the firm's track record – investments made by certain professionals or in certain industries – because those are representative of the next Fund's strategy.

Such track records are "synthetic". They lack true Net MOICs and Net IRRs because investments are not made in a dedicated Fund with Fees & Expenses and Carried Interest. Regulatory guidelines, however, insist that any marketing materials providing Gross Returns must also display Net Returns with equal prominence. Consequently, synthetic Net Returns are generated, one way or another, with little or no consistency in methodologies or assumptions across different GPs and Funds.

A GP might divide a 12-year track record into two groups to represent two Funds. Applying appropriate *F&E* and *CI* may show that Fund I and Fund II each demonstrate 18% Net IRRs. Portfolio selection is arbitrary, however, and changing the first investment in Fund II to the last investment in Fund I might increase Fund I's Net IRR to 21% and lower Fund II's Net IRR to 8%. LPs should spend some time examining the impact of selection. Investments made from 2008 through 2020 could be modeled as two Fund vintages (2008 and 2014), three Fund vintages (2008, 2012, and 2016), four Fund vintages (2008, 2011, 2014, and 2017), etc. As each Synthetic Fund is ranked against vintage peers, analysts may find that different combinations tell quite different stories.

Consider a Synthetic Fund modeled from nine private company investments made across three different committed Funds. The analyst has access to the track record in Table 14.6 and the last financial statement for all three Funds, but no access to the underlying Gross or Net cash flows. How could the Synthetic Fund's Net MOIC and Net IRR be determined?

Table 14.6 Synthetic Fund returns

	Fund	**Capital Invested**	**Cash Realized**	**Current Value**	**Total Value**	**Gross MOIC**	**Gross IRR**
PapaCo	Fund I	$16.2	$46.3	$0.0	$46.3	2.86x	44.1%
QuebecCo	Fund I	18.2	25.4	0.0	25.4	1.40x	14.2%
RomeoCo	Fund I	25.4	56.7	0.0	56.7	2.23x	17.4%
SierraCo	Fund II	26.3	20.5	0.0	20.5	0.78x	-5.5%
TangoCo	Fund II	20.7	73.4	0.0	73.4	3.55x	52.5%
UniformCo	Fund II	16.4	37.8	0.0	37.8	2.30x	41.8%
VictorCo	Fund III	18.7	31.7	0.0	31.7	1.70x	69.5%
WhiskeyCo	Fund III	25.7	0.0	43.7	43.7	1.70x	16.4%
XrayCo	Fund III	15.5	0.0	16.3	16.3	1.05x	1.6%
Total Gross Returns		$183.1	$291.8	$60.0	$351.8	1.92x	24.1%

Using the financial statements to divide each Fund's *F&E* by *GIC* might provide %*F* values of 11.0% for Fund I, 15.0% for Fund II, and 6.0% for Fund III. Such variance in %*F* is not uncommon, even for Funds with identical fee structures and terms. Early in a Fund's life, capital may be deployed faster than Management Fees are collected, resulting in lower $\%F$ values. Further, the ability or willingness of GPs to collect transaction or monitoring fees can change with time and circumstance. This influences Management Fee Offsets and, therefore, $\%F$.

The first step is to determine appropriate $F\&E$ and $\%F$ values for the Synthetic Fund. Total $F\&E$ could be modeled on each investment's pro-rata share of actual $F\&E$ in its original Fund. Total $F\&E$ would then be determined from each investment's Capital Invested and each Fund's $\%F$, as follows:

$$F\&E = 11.0\% \cdot (\$16.2 + \$18.2 + \$25.4) + 15.0\% \cdot (\$26.3 + \$20.7 + \$16.4) + 6.0\% \cdot (\$18.7 + \$25.7 + \$15.5) = \$19.7$$

Dividing $F\&E$ by total Capital Invested would provide the Synthetic Fund's $\%F$:

$$\%F = F\&E/GIC = \$19.7/\$183.1 = 10.8\%$$

Then, Net MOIC can be determined by:

$$X_{NET} = X_{GROSS} \cdot \frac{1}{1+\%F} \cdot MIN\left[1, 1 - \%C + \%C \cdot \frac{1+\%F}{X_{GROSS}}\right]$$

$$= 1.921x \cdot \frac{1}{1.108x} \cdot MIN\left[1, 1 - 20\% + 20\% \cdot \frac{1.108x}{1.921x}\right] = 1.588x$$

Net IRR for the Synthetic Fund requires first determining the Effective Hold from the Gross MOIC and Gross IRR, as follows:

$$\Delta t' = \frac{\ln(MOIC_{GROSS})}{\ln(1 + IRR_{GROSS})} = \frac{\ln(1.921x)}{\ln(1 + 24.13\%)} = 3.02\ years$$

Then, Net IRR can be determined by:

$$IRR_{NET} = X_{NET}{}^{1/\Delta t'} - 1 = 1.588^{1/3.02} - 1 = 16.54\%$$

As soon as $\%F$ for the Synthetic Fund is determined, it is possible to complete the analysis in a single step, as follows:

$$IRR_{NET} = \left\{\frac{GR}{GIC} \cdot \frac{1}{1+\%F} \cdot MIN\left[1, 1 - \%C + \%C \cdot \frac{1+\%F}{GR/GIC}\right]\right\}^{\frac{\ln(1+IRR_{GROSS})}{\ln(MOIC_{GROSS})}} - 1$$

$$= \left\{\frac{\$351.8}{\$183.1} \cdot \frac{1}{1+10.8\%} \cdot MIN\left[1, 1 - 20\% + 20\% \cdot \frac{1+10.8\%}{\$351.8/\$183.1}\right]\right\}^{\frac{\ln(1+24.13\%)}{\ln(\$351.8/\$183.1)}} - 1$$

$$= 16.54\%$$

There is a practical elegance to this approach that may not be apparent from the rather ugly formula above. Because the framework attributes Fees & Expenses and Carried Interest to investments systematically, Net Returns are methodologically consistent for any combination of investments. In contrast, most cash flow-driven Net Return models start with a set of five or six assumptions that are subjective and only applicable to one particular combination of investments. For example, portfolio-level Net Returns for the Synthetic Fund might be determined as follows:

- Capital Invested of $181.3 requires a Fund with committed capital of $200.
- Management Fees are equal to 2% of committed capital for a certain number of years, and then lower amounts thereafter.
- Fund Expenses are driven by certain first year set-up costs, and then lower amounts thereafter.
- Management Fee Offsets may or may not reduce Fees & Expenses.
- Portfolio investment cash flows may be modeled on their actual inflow or outflow dates or different dates that are more representative of a single Fund.
- The resulting Carried Interest, equal to 20% of Net Profits, may be allocated to different investments, in different amounts, and at different times.

If the analyst also requires Net Returns for exited investments, then all assumptions must be revised because Capital Invested of $141.9 requires a Fund with committed capital of perhaps $155, rather than $200, and this impacts everything else. Then, any change in portfolio composition, like a WhiskeyCo exit, requires all assumptions to be revised again. In contrast, the value creation framework's assumptions are identical for any individual investment or combination of investments, whether it be the entire Fund, only exited investments, only deals in certain industries, etc.

The difference between the two approaches may be best illustrated by the footnotes or endnotes required to disclose them in GP marketing materials. The cash flow-driven model disclosure must speak to all six bullet points above and include additional detail about the amounts and timing of the various cash flow adjustments. In contrast, the value creation framework disclosure could be summarized as follows:

"Net Returns" are determined by increasing Capital Invested by each investment's pro-rata share of actual Fund I, Fund II, or Fund III Management Fees and Fund Expenses, and then reducing Cash Realized or Unrealized Value (as applicable) by each investment's pro-rata share of pooled Carried Interest (if any), which is equal to 20% Net Profits, defined as the greater of $0 or Total Value, less Capital Invested, Management Fees, and Fund Expenses."

This is an entirely accurate description of the framework, but some will take issue with the fact that it says nothing about cash flow timing. IRRs are so sensitive to timing that, surely, the footnote must reference the dates and relative magnitudes of *F&E* and *CI* adjustments, correct? Not according to the single step Net IRR formula above. The only input that has anything to do with timing is the Gross IRR, which is buried deep within the formula because it is a variable within a natural logarithm, which is the numerator of a fraction, which is the exponent of the Net MOIC. The next section explores the mechanics of what is happening within this complicated formula.

14.8. Real IRR versus Effective IRR

After 196 pages and over 200 value creation formulas, it must be admitted that none of the 50+ value creation IRR formulas produce a real IRR. As noted in Section 5.2, IRR is technically the discount rate that makes the present value of all investment cash flows equal to $0. Value creation IRRs are not the same as these ubiquitous cash flow-driven measurements of private equity performance. However, they provide estimates that are quite close to what a real IRR should be, and they require much less effort, and much less data, than cash flow-driven IRR models. Value creation IRRs are Effective IRRs, which may be defined as:

"Compound Annual Growth Rates (CAGRs) measured over Effective Holds ($\Delta t'$), which are dollar-weighted durations that attempt to capture the impact of all capital inflows and outflows occurring during a holding period."

In defense of the Effective IRR, it might be noted that real IRRs are not without their complications. Positive portfolio IRRs driven by early winners can be insensitive to later cash flows, while negative portfolio IRRs can be hyper-sensitive to later cash flows. Consequently, various techniques have been developed to determine whether real IRRs are meaningful. The time-zero IRR evaluates a portfolio's return as if the first cash flow for every investment occurs on the same day. The neutral-weighted IRR evaluates a portfolio's return as if equal amounts of capital are invested in every company. The time-zero-neutral-weighted IRR includes both adjustments. If real IRRs were entirely conclusive, these and other IRR validation techniques would not exist.

It should also be noted that Effective IRR is not the first performance metric to masquerade as a real IRR. Private equity benchmarks provided by Cambridge Associates, Preqin, and Pitchbook include average IRRs and weighted average IRRs for each vintage year. Like the Effective IRR, these are functions of IRRs, but not IRRs themselves. While Effective IRRs are not the same as real IRRs, they are usually acceptable alternatives considering:

- Real IRR measurements are not always meaningful.
- LPs rely on other benchmarks and measures that are not real IRRs.
- Effective IRRs are compatible with all the value creation formulas, and they provide more angles of investigation than the cash flow-driven models.
- Effective IRRs are usually quite close to real IRRs, and they can be completed with a fraction of the data in a fraction of the time.

How close are Effective IRRs to real IRRs? The latter can be a moving target, so it is difficult to generalize the spread. Effective Net IRR formulas are only concerned with Gross MOIC, Gross IRR, *%C*, and *%F*. Two different Funds may be identical in this regard but have different real Net IRRs because of the timing of Carried Interest deductions or other Fund-level adjustments that are not considered by the formulas above.

For example, in Table 14.4 the Reported Fund's Net IRR of 19.0% is driven by an assumption that $8.3 of Carried Interest is deducted at the end of the hold. If the GP instead distributes half of the Carried Interest 18 months earlier at the AlphaCo exit, Net IRR would be 18.4% but the Gross MOIC, Gross IRR, *%C*, and *%F* would be no different. To the Effective Net IRR, both scenarios look like 18.2%, so the variance between Effective IRR and real IRR could be characterized as either 0.8% or 0.2%.

The Synthetic Fund in the prior section provides an opportunity to experiment with cash flows and better understand the inner workings of the value creation framework's Effective IRR. Gross cash flows for the Synthetic Fund are as follows:

Table 14.7 Synthetic Fund underlying Gross cash flows

	Papa	Queb	Rome	Sier	Tang	Unif	Vict	Whis	Xray	Total
Dec-10	-16.2									-16.2
Jun-11		-18.2								-18.2
Dec-11			-25.4							-25.4
Jun-12				-26.3						-26.3
Dec-12	15.4				-20.7					-5.3
Jun-13						-16.4				-16.4
Dec-13	15.4	25.4					-18.7			22.1
Jun-14								-25.7		-25.7
Dec-14	15.4					12.6	31.7		-15.5	44.2
Jun-15										-
Dec-15					73.4	12.6				86.0
Jun-16										-
Dec-16			56.7	20.5		12.6				89.8
Jun-17										-
Dec-17								43.7	16.3	60.0
IRR	44.1%	14.2%	17.4%	-5.5%	52.5%	41.8%	69.5%	16.4%	1.6%	24.1%

The Gross Returns and Fees & Expenses described in Section 14.7 provide the necessary details to calculate Net MOICs and Net IRRs for the individual investments and Synthetic Fund Total.

Table 14.8 Synthetic Fund Effective Net Returns

	Papa	Queb	Rome	Sier	Tang	Unif	Vict	Whis	Xray	Total
Gross MOIC	2.86x	1.40x	2.23x	0.78x	3.55x	2.30x	1.70x	1.70x	1.05x	1.921x
χ_{FEES}	1.11x	1.11x	1.11x	1.15x	1.15x	1.15x	1.06x	1.06x	1.06x	1.108x
χ_{CARRY}	1.14x	1.04x	1.11x	1.00x	1.16x	1.11x	1.08x	1.08x	1.00x	1.093x
Net MOIC	1.79x	1.04x	1.47x	0.59x	2.01x	1.41x	1.29x	1.29x	0.93x	1.558x
$\Delta t'$	2.88	2.51	5.01	4.51	3.00	2.39	1.00	3.51	3.00	3.02
Net IRR	32.8%	7.7%	12.6%	-8.3%	38.7%	28.0%	47.9%	11.9%	-0.3%	16.54%

Since Net MOIC is Gross MOIC divided by χ_{FEES} and χ_{CARRY}, the Net MOIC formula may be expressed:

$$\chi_{NET} = \chi_{GROSS} \cdot \frac{1}{\chi_{FEES}} \cdot \frac{1}{\chi_{CARRY}}$$

$$= \frac{GR}{GIC} \cdot \frac{1}{\chi_{FEES}} \cdot \frac{1}{\chi_{CARRY}} = \frac{GR/\chi_{CARRY}}{GIC \cdot \chi_{FEES}}$$

Effectively, Gross MOIC is transformed into Net MOIC by dividing *GR* (sum of all outflows) by χ_{CARRY} and multiplying *GIC* (sum of all inflows) by χ_{FEES}. Such adjustments to inflow and outflow totals can also be made to individual inflows and outflows. This provides an identical Net MOIC and generates a set of Net cash flows that can form the basis of a cash flow-driven Net IRR model, as in Table 14.9.

Table 14.9 Synthetic Fund underlying Net cash flows

	Papa	Queb	Rome	Sier	Tang	Unif	Vict	Whis	Xray	Total
Dec-10	-18.0									-18.0
Jun-11		-20.2								-20.2
Dec-11			-28.2							-28.2
Jun-12				-30.3						-30.3
Dec-12	14.1				-23.8					-9.7
Jun-13						-18.9				-18.9
Dec-13	14.1	23.2					-19.8			17.5
Jun-14								-27.2		-27.2
Dec-14	14.1					11.5	29.0		-16.5	38.2
Jun-15										-
Dec-15					67.2	11.5				78.7
Jun-16										-
Dec-16			51.9	18.7		11.5				82.2
Jun-17										-
Dec-17								40.0	14.9	54.9
IRR	34.4%	5.7%	13.0%	-10.1%	41.3%	28.4%	46.4%	11.6%	-3.2%	16.31%

In the model above, every Gross inflow (portfolio investment) is multiplied by the deal-by-deal χ_{FEES} value in Table 14.8, and every Gross outflow (realization and unrealized value) is divided by the pooled χ_{CARRY} value of 1.093x. This provides a cash flow-driven Net IRR of 16.31%, which is within 0.23% of the Effective Net IRR of 16.54%. The approach is identical to what is described by the footnote or endnote disclosure in Section 14.7:

"Net Returns" are determined by increasing Capital Invested by each investment's pro-rata share of actual Fund I, Fund II, or Fund III Management Fees and Fund Expenses, and then reducing Cash Realized or Unrealized Value (as applicable) by each investment's pro-rata share of pooled Carried Interest (if any), which is equal to 20% Net Profits, defined as the greater of $0 or Total Value, less Capital Invested, Management Fees, and Fund Expenses."

The disclosure requires no details about the timing of *F&E* and *CI* adjustments because no new inflows or outflows are added to the model. Instead, every inflow is increased by a factor (χ_{FEES}) while every outflow is decreased by a factor ($1/\chi_{CARRY}$). This systematically transforms Gross cash flows into Net cash flows and provides a Net Return model that is methodologically consistent for any combination of investments.

Analysts have the option of using either the pooled or deal-by-deal measures for χ_{FEES} and χ_{CARRY}. Deal-by-deal χ_{FEES} values are used in Table 14.9. Alternatively, the pooled χ_{FEES} value of 1.108x can be applied to each company. This does not change total *F&E* or *CI*, so it has no impact on Net MOIC. However, it causes minor shifts in cash flow magnitudes that change Net IRR to 16.43%. This is even closer to the Effective Net IRR of 16.54%, which also uses pooled values for χ_{FEES} and χ_{CARRY}.

The Effective Net IRR is a streamlined version of the model above, where Net cash flows are determined from Gross cash flows and pooled χ_{FEES} and χ_{CARRY} values. The Synthetic Fund's Gross MOIC and Gross IRR provide an Effective Hold of 3.020297... years, so the Effective Net IRR model takes the 22 cash flows occurring between 12/31/2010 and 12/31/2017 and effectively reduces them to two cash flows: one inflow on the morning of 12/27/2012 and one outflow on the afternoon of 01/03/2016. As illustrated in Table 14.10, this provides an Effective Gross IRR of 24.14% and an Effective Net IRR of 16.54%, which are almost identical to the cash flow-driven models above.

Table 14.10 Illustration of the inner workings of the Effective IRR

	Gross Total	Factor	Net Total
12/27/2012, 7:06 AM	-183.1	1.108x	-202.8
01/03/2016, 4:43 PM	351.8	1/1.093x	322.0
IRR	24.14%		16.54%

The value creation framework is the most efficient way to calculate robust and reliable Net IRRs for individual investments and portfolios. The Effective Net IRR does not require time series data or complicated cash flow models, and it is built from input variables like Gross MOIC and Gross IRR that are generally available in Fund reporting and GP marketing materials. The only drawback of the framework is that it is uncommon. Many GPs, LPs, and regulators are unfamiliar with it, and they may be skeptical about the lack of timing details for Fees & Expenses and Carried Interest adjustments.

If a traditional cash flow-driven Net IRR model is required, it would be best to employ an approach like Table 14.9, where Fees & Expenses and Carried Interest are applied to the Gross cash flows systematically, rather than an inflexible model that requires making various subjective assumptions about Fund sizes and cash flow adjustments. The systematic application of χ_{FEES} and χ_{CARRY} makes models more scalable and better able to accommodate changes in portfolio composition, while providing a methodologically consistent approach for Funds, individual investments, and combinations of investments within or across Funds.

14.9. Summary

Starting with the Gross Returns for a private company or portfolio and a reasonable estimate for Management Fees and Fund Expenses, the value creation framework provides a robust and reliable way to determine Net Returns. The approach models Fees & Expenses and Carried Interest as value drivers that "bridge the gap" between Gross Returns and Net Returns.

The framework makes systematic adjustments to Gross capital inflows and outflows, which eliminates much of the subjective guesswork that is common with traditional Net Return models. Its formulas are methodologically consistent for any individual investment or combination of investments, whether it be the entire Fund, only exited investments, only deals in certain industries, etc. This allows portfolios to be broken into components, rearranged, and reconstructed into new Funds with no loss of data integrity. Like the return measurements and value drivers described in prior chapters, results are mathematically rigorous and can express returns in the form of absolute ($) value creation, relative (%) value creation, equity return multiples or multipliers, and value creation IRRs.

The value creation framework can be used to transform a set of portfolio Gross cash flows into a set of Fund-like Net cash flows, which provides a familiar model for calculating Net IRRs. The Effective Net IRR formulas skip the cash flow models and provide very good Net IRR estimates using the Effective Hold ($\Delta t'$) and input variables that are generally available in Fund reporting and GP marketing materials.

14.10. Optional: Net Return-based Value Creation

Relative return measurements like relative (%) value creation, equity return multiples, and value creation IRRs generally use the Gross Return as a reference point. For example, relative (%) value creation driven by EBITDA Growth is:

$$v_{EG} = V_{EG}/\Delta FEqV = V_{EG}/V_{GROSS}$$

The expression's denominator is defined as:

$$V_{GROSS} = \Delta FEqV = FEqV_2 - FEqV_1 = GR - GIC$$

This is typically the best way to evaluate individual investment and portfolio returns. While Net Returns are ultimately the most important numbers to LPs, a GP's proficiency at its "day job", which is the buying and selling of private companies, is best measured by Gross Return, before any deduction for Fees & Expenses and Carried Interest.

However, in certain situations, it may be helpful to use Net Return as a reference point. This is demonstrated in the "Stakeholder Model" of value creation in Section 1.7, where value drivers are aggregated as follows:

- Operational Impact is Market Share Growth, EBITDA Margin Expansion, and Intrinsic Multiple Expansion.
- Financing Impact is Leverage Effect, Fund Ownership Impact, Fees & Expenses, and Carried Interest.
- Market Impact is Addressable Market Growth and Market Multiple Expansion.

Table 14.11 Stakeholder value creation model

	Absolute ($) Net Value Creation	Relative (%) Net Value Creation	Net IRR
Operational Impact	$22.8	73.6%	9.4%
Financing Impact	-$6.4	-20.5%	-2.5%
Market Impact	$14.5	46.9%	5.9%
Net Total	$31.0	100.0%	13.0%

In this model, v_{EG} would be defined as V_{EG}/V_{NET}, where:

$$V_{NET} = FEqV_2 - FEqV_1 - F\&E - CI = NR - NIC$$

When solving for Gross Return-based value drivers, *F&E* and *CI* are positive or neutral values (≥$0, ≥0%, or ≥1.0x) that work to increase a Net Return to the level of Gross Return. The equity return multiplier expression is:

$$\chi_{GROSS} = \chi_{NET} \cdot \chi_{FEES} \cdot \chi_{CARRY}$$

When solving for Net Return-based value drivers, *F&E* and *CI* must be negative or neutral values (≤$0, ≤0%, or ≤1.0x) that work to decrease a Gross Return to the level of Net Return. In this framework, the *F&E* and *CI* equity return multipliers are their reciprocals, and any expression for χ_{GROSS} can be substituted into the following formula:

$$\begin{aligned} \chi_{NET} &= \frac{1}{\chi_{FEES}} \cdot \frac{1}{\chi_{CARRY}} \cdot \chi_{GROSS} \\ &= \frac{1}{\chi_{FEES}} \cdot \frac{1}{\chi_{CARRY}} \cdot \chi_{EG} \cdot \chi_{ME} \cdot \chi_{LEV} \cdot \chi_{FO} \end{aligned}$$

The Net Return-based expression for value creation IRRs can be determined from equity return multipliers:

$$\begin{aligned} (1 + IRR_{NET}) = (1 + IRR_{FEES}) \cdot (1 + IRR_{CARRY}) \cdot (1 + IRR_{EG}) \cdot (1 + IRR_{ME}) \\ \cdot (1 + IRR_{LEV}) \cdot (1 + IRR_{FO}) \end{aligned}$$

$$\chi_{NET}{}^{1/\Delta t} = \left(\frac{1}{\chi_{FEES}}\right)^{1/\Delta t} \cdot \left(\frac{1}{\chi_{CARRY}}\right)^{1/\Delta t} \cdot \chi_{EG}{}^{1/\Delta t} \cdot \chi_{ME}{}^{1/\Delta t} \cdot \chi_{LEV}{}^{1/\Delta t} \cdot \chi_{FO}{}^{1/\Delta t}$$

The Net Return-based expression for relative (%) value creation can be determined by taking the natural logarithm of both sides of the equity return multiplier expression and dividing by $\ln(\chi_{NET})$:

$$\begin{aligned} &\frac{\ln(\chi_{NET})}{\ln(\chi_{NET})} = 100\% \\ &= \frac{\ln\left(\frac{1}{\chi_{FEES}} \cdot \frac{1}{\chi_{CARRY}} \cdot \chi_{EG} \cdot \chi_{ME} \cdot \chi_{LEV} \cdot \chi_{FO}\right)}{\ln(\chi_{NET})} = 100\% \\ &= \frac{\ln(1/\chi_{FEES})}{\ln(\chi_{NET})} + \frac{\ln(1/\chi_{CARRY})}{\ln(\chi_{NET})} + \frac{\ln(\chi_{EG})}{\ln(\chi_{NET})} + \frac{\ln(\chi_{ME})}{\ln(\chi_{NET})} + \frac{\ln(\chi_{LEV})}{\ln(\chi_{NET})} + \frac{\ln(\chi_{FO})}{\ln(\chi_{NET})} \\ &= 100\% \\ &= v_{FEES} + v_{CARRY} + v_{EG} + v_{ME} + v_{LEV} + v_{FO} = 100\% \end{aligned}$$

The Net Return-based expression for absolute ($) value creation can be determined from relative (%) value creation:

$$\begin{aligned} V_{NET} &= V_{FEES} + V_{CARRY} + V_{EG} + V_{ME} + V_{LEV} + V_{FO} \\ &= v_{FEES} \cdot V_{NET} + v_{CARRY} \cdot V_{NET} + v_{EG} \cdot V_{NET} + v_{ME} \cdot V_{NET} + v_{LEV} \cdot V_{NET} \\ &\quad + v_{FO} \cdot V_{NET} \end{aligned}$$

Such Net Return-based value creation expressions may prove useful in certain circumstances or specialized situations, but the Gross Return-based expressions are more common and generally provide a better representation of company-level value creation activities over a holding period.

15. Conclusion and Next Steps

Private Equity Value Creation Analysis describes more than two dozen ways to interpret and discuss private company and portfolio returns. The analytical framework is based on the principles and best practices outlined in Chapter 3, which provide value creation models that:

- Are mathematically rigorous, meaning that every value driver can be calculated directly, and values can be converted back and forth between the various models and measures with no loss of data integrity
- Rely on accessible, practical, and non-volatile input variables that are generally available in Fund reporting and GP marketing materials
- Function over a wide range of company, industry, and market conditions
- Provide stable and intuitive results that are easy to explain and understand

15.1. Summary of Approach

Volume I started with the most common value creation components, including company-level EBITDA Growth, Multiple Expansion, and the impact of debt or leverage. It then introduced a framework for evaluating returns for specific shareholders, like Funds or GPs. The subsequent chapters showed how changes in company P&Ls impact Fund-level returns, before exploring how broader market or industry forces affect these various P&L and balance sheet components. The last chapter demonstrated how the value creation framework can calculate Net Returns for private company investments and portfolios, including Net IRRs, without time series data or complicated cash flow models.

Another reasonable approach would have been to start with the Fund-level Gross Return and deconstruct the value creation components as illustrated in Figure 15.1. Gross Return is at the top of the figure because it represents the fundamental measurement of GP performance, the buying and selling of private companies. It can be broken into components of Net Return, Fees & Expenses, and Carried Interest on the left or the various P&L, balance sheet, and market components on the right.

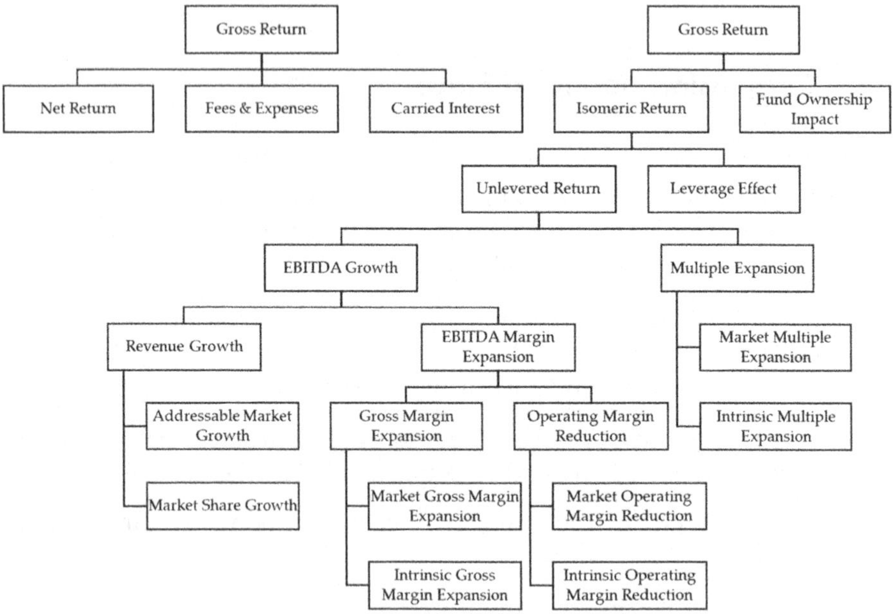

Figure 15.1 Value creation component hierarchy

On the right side, the first step rationalizes equity on the company capitalization table by separating Isomeric Return and Fund Ownership Impact. The second step addresses equity and debt on the company balance sheet to derive Unlevered Return and the Leverage Effect. The third step considers enterprise valuation and performance to provide Multiple Expansion and EBITDA Growth. Everything downstream from EBITDA Growth is tied to company P&L metrics, including Revenue Growth, EBITDA Margin Expansion, Gross Margin Expansion, and Operating Margin Reduction.

With the P&L and balance sheet components defined, market proxies can be used to isolate market-driven return components. On the valuation side,

Multiple Expansion can be expressed as Market Multiple Expansion and Intrinsic Multiple Expansion. On the P&L side, Revenue Growth can be expressed as Addressable Market Growth and Market Share Growth. Market Multiple Expansion and Addressable Market Growth are the most reliable market-driven components, and *Private Equity Value Creation Analysis* defines their combination as the Market-driven Return. However, if reasonable market proxies can be found, the market's impact on other components like EBITDA Margin Expansion, Gross Margin Expansion, and Operating Margin Reduction can also be measured.

15.2. Next Steps

As suggested by the subtitle, *Volume I: Theory*, this volume is principally dedicated to deriving and describing the various value creation formulas and making the case that the analytical framework described by *Private Equity Value Creation Analysis* is robust, reliable, and easier to use than many alternative techniques. With the theoretical and mathematical framework fully described, the next step is to focus on application. Errors driven by bad value creation input variables can be worse than those caused by bad value creation formulas. *Volume II: Applications* will provide more examples and case studies of analyses across various transaction types and industries. It will also help analysts gather and correctly format value creation input variables for numerous scenarios that are likely to be encountered in the real world.

Many of the topics covered in *Volume I* and *Volume II* are available on the author's private equity mathematics website, AUXILIA Mathematica. The site includes a private forum and online calculators that perform many of the calculations described in preceding chapters. Supporting spreadsheet files are available to purchasers of this book and include value creation input data, formulas, and graphs, which could be used as templates for their own analyses. Registration is free and allows users to comment and communicate with each other anonymously, while keeping spammers, scammers, and other internet ne'er-do-wells off the site. Sign up below:

https://auxiliamath.com/register

15.3. Conclusion

Private equity is a global multi-trillion-dollar asset class that is devoted to the buying and selling of private companies. Every year, several hundred billion in private capital flows from institutional investors to Fund managers, and then into private company investments, where it hopes to generate outsized equity returns and then make the return trip. While capital generally flows toward the best historical track records, it is also influenced by the quality of the stories that Fund managers and institutional investors tell each other about why companies succeed, why others struggle, and how broader market trends impact portfolios. Value creation analysis can add quantitative insights to these qualitative discussions, making Fund managers better story tellers and helping institutional investors validate whether these narratives are supported by the numbers.

16. Preview of Volume II: Applications

As noted in the prior chapter, errors driven by bad value creation input variables can be worse than those caused by bad value creation formulas. This volume's companion, *Volume II: Applications*, will help analysts gather and correctly format input variables for numerous scenarios that are likely to be encountered in the real world. It will also expand the universe of analysis candidates to include additional transaction types, investment stages, and industries. Many of the topics covered in *Volume I* and *Volume II* are available on the author's private equity mathematics website, AUXILIA Mathematica (www.auxiliamath.com).

16.1. The LP Request

Within a few days of any significant realization, GPs begin to field questions from LPs about P&L and capital structure metrics at exit. A value creation analysis is probably underway if they ask for details like exit price, net debt, EBITDA, and exit multiple. Requests for the same metrics at entry make it a certainty. Since such analyses are used to evaluate a GP's "operational impact" or investment manager "skill", it is worth pausing to examine those numbers before fulfilling the request.

These input variables suggest that an LP is measuring EBITDA Growth (V_{EG}), Multiple Expansion (V_{ME}), and a balance sheet component which may be either Cash Flow Generation (V_{CF}) or the Leverage Effect (V_{LEV}). All else being equal, GPs prefer that more value be assigned to EBITDA Growth, because valuations are often attributed to broader market forces ("luck") and balance sheet components may be offset by the risks of owning companies with more debt. It turns out that the way numbers are presented can strongly influence how much value creation is attributed to each component.

16.2. Value Creation Input Variables

YankeeCo is a solid investment that more than doubles shareholder equity. According to the GP's investment committee memorandum, the entry purchase price is \$100, which represents 5.0x an entry EBITDA of \$20. It is funded with \$20 of net debt, which is paid down to \$6 at exit. The exit price is \$180, which represents 6.0x an exit EBITDA of \$30. This suggests the following value creation input variables:

Table 16.1. YankeeCo value creation input variables

	t_1	t_2	ΔX	$\overline{X}$	X_2/X_1
TEV	\$100.0	\$180.0	+\$80.0	\$140.0	1.80x
TEqV	\$80.0	\$174.0	+\$94.0	-	2.25x
ND	\$20.0	\$6.0	-\$14.0	-	n/a
ER	80.0%	96.7%	+16.7%	88.3%	1.25x
E	\$20.0	\$30.0	+\$10.0	\$25.0	1.50x
M	5.00x	6.00x	+1.00x	5.50x	1.20x

The metrics above are suitable for calculating value creation with any of the three models described in *Volume I*. This chapter focuses on the Conventional Model because it best illustrates equity value changes. Its slightly higher EBITDA Growth and Multiple Expansion components, driven by the Conventional Model's tendency to understate the impact of leverage, increases its sensitivity to the value creation input variables.

According to the formulas in Chapter 4, YankeeCo generates absolute (\$) value creation of \$55 for V_{EG}, \$25 for V_{ME}, and \$14 for V_{CF}, which corresponds to relative (%) value creation of 58.5% for v_{EG}, 26.6% for v_{ME}, and 14.9% for v_{CF}.

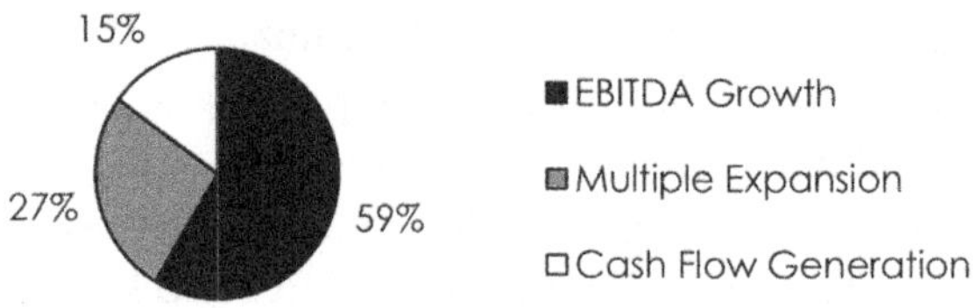

Figure 16.1 YankeeCo initial analysis

Table 16.1 reflects the way that private company entry and exit metrics are commonly reported and discussed, but it is not the best format for value creation analysis. Such metrics are often estimates that reflect how GPs think about or negotiate transactions. They do not necessarily tie to actual capital movements between buyers and sellers, unless there are corrections to account for various items that arise in a sale process. This chapter shows that uncorrected input variables tend to move value creation away from EBITDA Growth and into Multiple Expansion, so it is a mistake for GPs to neglect them. Examples of five such "mistakes" are demonstrated in the following sections.

16.3. Mistake #1: Overstating Entry EBITDA

Deal professionals love low entry multiples. There are no known exceptions to this rule. When LPs request portfolio company entry multiples and provide no context or direction about the value's format, GPs do what seems natural and provide the lowest possible number that can be rationally defended. The entry multiple's denominator is usually EBITDA or revenue, and higher values make it look like the GP structures a better deal and acquires more performance for the stated price. However, higher entry EBITDAs also suggest less growth over the hold and, therefore, less value creation attributed to EBITDA Growth.

Entry EBITDA might be overstated due to timing discrepancies. YankeeCo's E_1 of \$20 may be a full year estimate, whereas the transaction closes six months earlier, on June 30. If YankeeCo is growing at entry, which is usually the case for companies that successfully trade, TTM June 30 performance is lower than \$20 and ΔE must be higher. If YankeeCo's E_1 is instead \$19, the input variables are:

Table 16.2. YankeeCo value creation input variables after entry EBITDA correction

	t_1	t_2	ΔX	$\overline{X}$	X_2/X_1
TEV	\$100.0	\$180.0	+\$80.0	\$140.0	1.80x
TEqV	\$80.0	\$174.0	+\$94.0	-	2.25x
ND	\$20.0	\$6.0	-\$14.0	-	n/a
ER	80.0%	96.7%	+16.7%	88.3%	1.21x
E	\$19.0	\$30.0	+\$11.0	\$24.5	1.58x
M	5.26x	6.00x	+0.74x	5.63x	1.14x

According to the Conventional Model, the increase in ΔE to \$11 and the decrease in ΔM to 0.74x shifts \$6.9 of value out of Multiple Expansion and into EBITDA Growth, causing a 14.8% swing in relative (%) value creation.

Figure 16.2 YankeeCo analysis after entry EBITDA correction

There is nothing wrong with GPs citing full year estimates in their investment materials. They can argue that they "bought off" the higher forward numbers, which are often highly certain and key assumptions of their underwriting cases. However, adjustments on the order of 5% are common in private equity, and the example above shows that they can be impactful. Aspirational adjustments for items like anticipated synergies may overstate entry EBITDAs by similar amounts. From a value creation perspective, GPs are usually best off excluding such adjustments and using TTM periods that end as close as possible to the transaction date.

16.4. Mistake #2: Neglecting Entry Fees, Expenses, and Adjustments

The entry multiple's numerator is entry price, but the headline purchase prices commonly discussed are not the best value creation input variables. These are usually pre-money valuations that reflect what sellers receive at close. The value creation starting point is a post-money valuation, or what buyers own on the first day of their hold. Entry prices should include buyer transaction fees and expenses and other adjustments for working capital, retained cash, acquired liabilities, seller notes, etc., which impact both $TEqV$ and TEV. If the YankeeCo acquisition includes buyer fees and expenses of $3, total equity required to fund the transaction is $83 (rather than $80) and entry price is $103 (rather than $100), providing the following input variables:

Table 16.3. YankeeCo value creation input variables after entry price correction

	t_1	t_2	ΔX	$\bar{X}$	X_2/X_1
TEV	$103.0	$180.0	+$77.0	$141.5	1.75x
$TEqV$	$83.0	$174.0	+$91.0	-	2.10x
ND	$20.0	$6.0	-$14.0	-	n/a
ER	80.6%	96.7%	+16.1%	88.6%	1.20x
E	$19.0	$30.0	+$11.0	$24.5	1.58x
M	5.42x	6.00x	+0.58x	5.71x	1.11x

According to the Conventional Model, the increase in $\overline{M}$ to 5.71x enhances EBITDA Growth by \$0.9 and the decrease in ΔM to 0.58x reduces Multiple Expansion by \$3.9, causing a 6.7% swing in relative (%) value creation. This demonstrates that more than 100% of any neglected entry fees and expenses go directly to Multiple Expansion and cause modest reductions in EBITDA Growth.

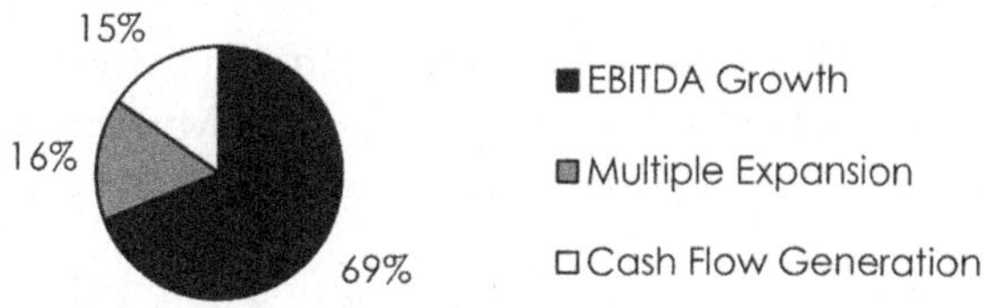

Figure 16.3 YankeeCo analysis after entry price correction

Several sections in *Volume II* focus on correctly formatting entry prices for numerous scenarios that are likely to be encountered in the real world. Different transaction situations require different adjustments to arrive at correct TEV values. Analysts may need to access buyer funds flow files and be aware of subsequent changes that would affect the value creation "starting point", like later working capital adjustments, planned equity sell downs to management or co-investors, or the replacement of short-term debt securities, like seller notes.

16.5. Mistake #3: Understating Exit EBITDA

Overstated exit multiples have the same impact as understated entry multiples, which is to shift value creation away from EBITDA Growth. They may be caused by understated exit EBITDAs, which can also be driven by timing discrepancies. When an LP requests exit metrics for a company sold on June 30, it is not uncommon for a deal team to provide the latest available or "approved" financials from company management. These may be TTM March 31 or the prior full year, which are usually lower than TTM June 30 values because most companies that successfully trade are growing at exit. Furthermore, the buyer may have "bought off" a higher, forward EBITDA (just as the seller did at entry in Section 16.3). If YankeeCo's E_2 is instead closer to \$31, the input variables are:

Table 16.4. YankeeCo value creation input variables after exit EBITDA correction

	t_1	t_2	ΔX	$\overline{X}$	X_2/X_1
TEV	\$103.0	\$180.0	+\$77.0	\$141.5	1.75x
TEqV	\$83.0	\$174.0	+\$91.0	-	2.10x
ND	\$20.0	\$6.0	-\$14.0	-	n/a
ER	80.6%	96.7%	+16.1%	88.6%	1.20x
E	\$19.0	\$31.0	+\$12.0	\$25.0	1.63x
M	5.42x	5.81x	+0.39x	5.61x	1.07x

According to the Conventional Model, the increase in ΔE to \$12 and the decrease in ΔM to 0.39x shifts \$4.5 of value out of Multiple Expansion and into EBITDA Growth, causing a 10.0% swing in relative (%) value creation.

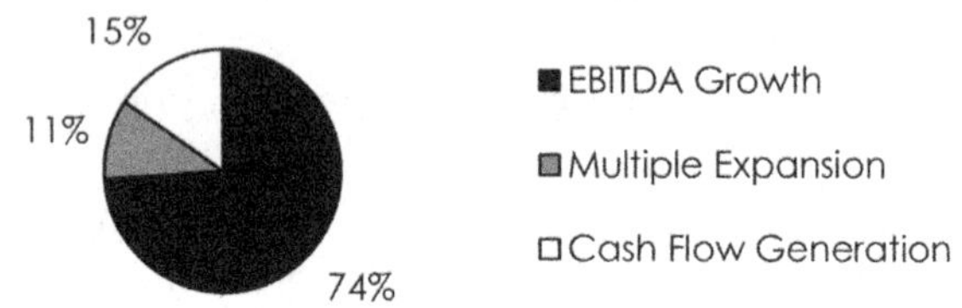

Figure 16.4 YankeeCo analysis after exit EBITDA correction

The understatement of exit P&L values usually has less to do with GP biases and more to do with a lack of attention. Deal professionals are less invested in exit metrics because their investment approaches and track records are not usually defined by them. Fatigue from extended sale processes or holding periods that last several years longer than plan may also play a role. Regardless, exit metrics can impact value creation results as much as entry metrics, so they require equal consideration.

From a value creation perspective, the best P&L ending points reflect what is "known" at the time of closing and represent a TTM period that ends as close as possible to the transaction date. Since there is often a lag in P&L reporting, estimates may be required. For example, a TTM June 30 exit EBITDA might be calculated from nine months of actual values through March 31 and budget or projected values for April, May, and June.

16.6. Mistake #4: Neglecting Exit Fees, Expenses, and Adjustments

Headline exit prices have the same problems as headline entry prices. They usually exclude the impact of seller transaction fees, expenses, and other adjustments. Neglecting them cause equity values to not match the proceeds delivered to selling shareholders and tends to cause value creation analyses to overstate the impact of Multiple Expansion. *Volume II* addresses exit adjustments in detail and describes a company, like YankeeCo, with a headline exit price of $180. In the example, the exit proceeds are reduced by $4.2 for seller transaction fees and expenses, $1.0 for delivering less than sufficient working capital to buyers, and $0.8 for other adjustments, including a lender pre-payment penalty and foreign tax liability. These adjustments reduce equity proceeds to $168 and exit price to $174, which provides the following input variables:

Table 16.5. YankeeCo value creation input variables after exit price correction

	t_1	t_2	ΔX	$\overline{X}$	X_2/X_1
TEV	$103.0	$174.0	+$71.0	$138.5	1.69x
TEqV	$83.0	$168.0	+$85.0	-	2.02x
ND	$20.0	$6.0	-$14.0	-	n/a
ER	80.6%	96.6%	+16.0%	88.6%	1.20x
E	$19.0	$31.0	+$12.0	$25.0	1.63x
M	5.42x	5.61x	+0.19x	5.52x	1.04x

According to the Conventional Model, the decrease in $\overline{M}$ to 5.52x reduces EBITDA Growth by $1.2, but this still represents a relative (%) increase in EBITDA Growth because the decrease in ΔM to 0.19x reduces Multiple Expansion by $4.8, causing an 8.8% swing in relative (%) value creation.

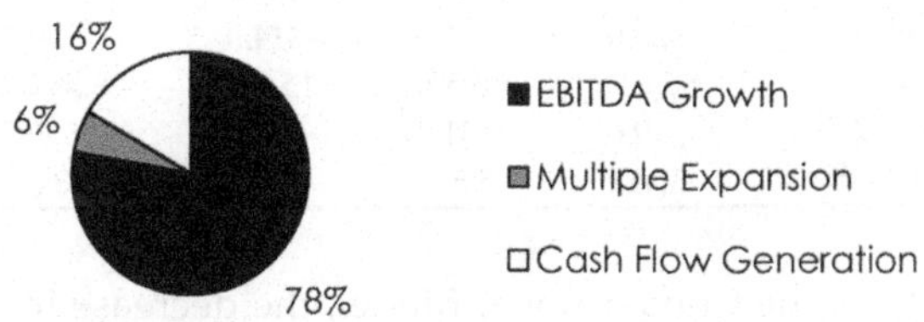

Figure 16.5 YankeeCo analysis after exit price correction

Headline purchase prices should never be used as value creation inputs, either at entry or at exit. They are basically reference points that help buyers and sellers to negotiate and keep track of dozens of price-related adjustments. Such adjustments must be reflected in exit price to ensure that equity values tie out to the capital flows among buyers, sellers, management, lenders, service providers, etc. This includes adjustments that take place after the transaction date, which may be months later in the case of working capital adjustments and years later in the case of escrow proceeds.

16.7. Mistake #5: Neglecting Holdback and Escrow Discounts

Holdbacks and escrows are established to protect buyers against undisclosed liabilities that may arise after closing. Some amount of sale proceeds is set aside for a number of months or years, to be released according to predetermined mechanisms defined in a purchase agreement. Sellers may eventually receive all, some, or none of the holdback or escrow proceeds, so they are generally held at a discount to face value until proceeds are ultimately released.

Assume that YankeeCo has a holdback of $1.0 related to a working capital target to be determined after closing and an escrow of $4.0 related to uncertainty around questionable customer contracts. The seller may apply a 50% discount to the escrow, but only a 20% discount to the holdback because it is shorter term and there is good visibility into working capital. The total discount of $2.2 reduces exit proceeds to $165.8 and exit price to $171.8. If the holdback is released in full and the 50% escrow discount is maintained, the total discount changes to $2.0, which provides realized and unrealized exit proceeds of $166.0, an exit price of $172.0, and the following input variables:

Table 16.6. YankeeCo value creation input variables after exit escrow correction

	t_1	t_2	ΔX	$\overline{X}$	X_2/X_1
TEV	$103.0	$172.0	+$69.0	$137.5	1.67x
TEqV	$83.0	$166.0	+$83.0	-	2.00x
ND	$20.0	$6.0	-$14.0	-	n/a
ER	80.6%	96.5%	+15.9%	88.5%	1.20x
E	$19.0	$31.0	+$12.0	$25.0	1.63x
M	5.42x	5.55x	+0.13x	5.48x	1.02x

According to the Conventional Model, the decrease in $\overline{M}$ to 5.48x reduces EBITDA Growth by $0.4, but this still represents a relative (%) increase in EBITDA Growth because the decrease in ΔM to 0.13x reduces Multiple Expansion by $1.6, causing a 3.2% swing in relative (%) value creation.

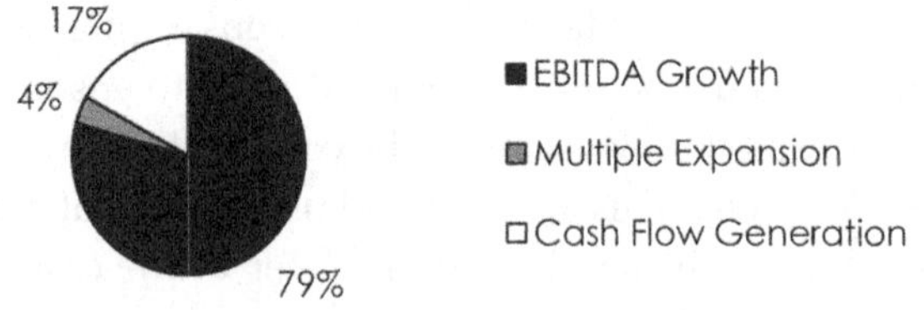

Figure 16.6 YankeeCo final analysis after exit escrow correction

This correction shows that price adjustments on the order of 1% or 2% can materially impact value creation results, and the final exit price may depend on events that occur long after closing. Input variables like TEV_2, $TEqV_2$, ER_2, and M_2 might continue to change until all holdbacks and escrows are released, causing the value creation components to fluctuate until the last unrealized dollar is recovered.

16.8. Summary

Correcting the five "mistakes" described in the prior sections does exactly what any GP would hope for, which is to shift significant value out of Multiple Expansion and into EBITDA Growth. Each moves value creation components only modestly, but they combine to dramatically alter the value creation story because each correction tends to shift value in the same direction. Multiple Expansion accounts for 26.6% of equity value change in the initial analysis. Every correction reduces Multiple Expansion until it accounts for only 3.8% of value creation in the final analysis.

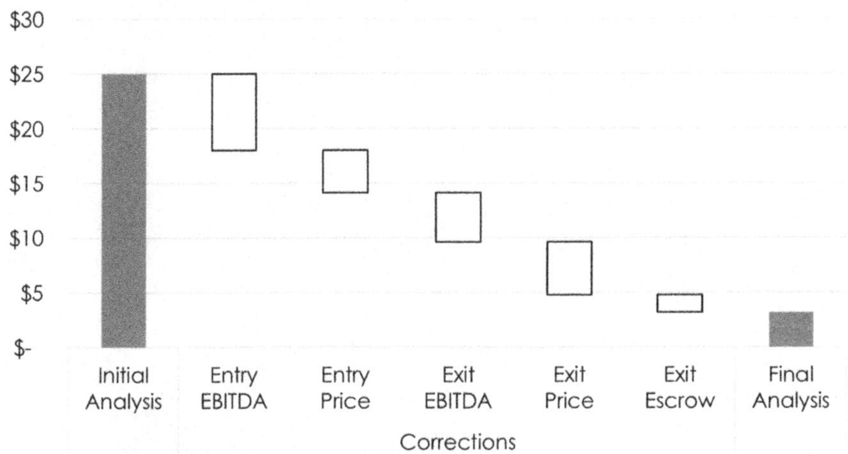

Figure 16.7 Impact of corrections on YankeeCo Multiple Expansion

When answering questions about YankeeCo's entry and exit metrics, the GP should provide the numbers in Table 16.6 rather than Table 16.1. If different headline purchase prices or transaction multiples have already been shared with LPs, the GP should simply note that the corrected metrics are more appropriate for value creation analyses because they account for transaction fees and expenses, as well as other closing adjustments. Such corrections are not made to "game" the value creation analysis, but rather to calculate more meaningful value creation components. Essentially, the five corrections ensure that: 1) TTM P&L variables line up with the entry and exit dates, and 2) entry and exit prices and equity values tie to the capital that flows back and forth between the stakeholder bank accounts. The tendency of corrections to shift value toward EBITDA Growth and the downstream P&L components that are associated with GP "operational impact" or investment manager "skill" is merely a convenient side-effect of performing a more accurate analysis.

References

Achleitner, A., and K. Lichtner, C. Diller. 2009. Value Creation in Private Equity. Capital Dynamics, the Center for Entrepreneurial and Financial Studies (CEFS) at the Technische Universität München.

Long, A., and C. Nickels. 1996. A Private Investment Benchmark. A paper presented at the University of Texas System, AIMR Conference on Venture Capital Investing. http://www.ww.w.alignmentcapital.com/pdfs/research/icm_aimr_benchmark_1996.pdf (accessed September 30, 2020).

Morris, Peter. 2014. Approach Private Equity's "Value Bridge" with Caution. *Financial World*, October/November 2014. https://ssrn.com/abstract=2563532 (accessed September 30, 2020).

PitchBook Data, Inc. 2020. Private Fund Strategies Report: 2019 Annual. https://files.pitchbook.com/website/files/pdf/PitchBook_2019_Annual_Private_Fund_Strategies_Report.pdf (accessed September 30, 2020).

PitchBook Data, Inc. 2020. US PE Middle Market Report: 2019 Annual. https://files.pitchbook.com/website/files/pdf/PitchBook_2019_Annual_US_PE_Middle_Market_Report.pdf (accessed September 30, 2020).

Preqin Ltd. 2020. Preqin Investor Outlook: Alternative Assets H1 2020. https://docs.preqin.com/reports/Preqin-Investor-Outlook-Alternative-Assets-H1-2020.pdf (accessed September 30, 2020).

S&P CapitalIQ. 2020. S&P 500 Software (Industry) (^SP500-451030). Retrieved August 11, 2020, from S&P CapitalIQ database.

US Bureau of Economic Analysis. 2020. Gross Output by Industry, Release Date: October 29, 2019. Retrieved August 12, 2020, from US Bureau of Economic Analysis. https://apps.bea.gov/iTable/iTable.cfm?reqid=150&step=2&isuri=1&categories=gdpxind.

White, B., and M. Prahl, C. Zeisberger. 2016. Value Creation 2.0. INSEAD Global Private Equity Initiative (GPEI). https://www.insead.edu/sites/default/files/assets/dept/centres/emi/docs/value-creation-2-0.pdf (accessed September 30, 2020).

Index

Made in the USA
Las Vegas, NV
17 January 2024

84495205R00134